THE DISAPPEARANCE OF
TELECOMMUNICATIONS

Books of Related Interest from IEEE Press

INTEGRATED TELECOMMUNICATIONS MANAGEMENT SOLUTIONS
A book in the IEEE Press Series on Network Management
Graham Chen and Qinzheng Kong
2000   Hardcover   288 pp   IEEE Order No. PC5795   ISBN 0-7803-5353-6

SECURITY FOR TELECOMMUNICATIONS NETWORK MANAGEMENT
A book in the IEEE Press Series on Network Management
Moshe Rozenblit
2000   Hardcover   320 pp   IEEE Order No. PC5393   ISBN 0-7803-3490-6

TELECOMMUNICATIONS NETWORK MANAGEMENT: Technologies and Implementations
A book in the IEEE Press Series on Network Management
Salah Aidarous and Thomas Plevyak
1998   Hardcover   352 pp   IEEE Order No. PC5711   ISBN 0-7803-3454-X

TELECOMMUNICATIONS NETWORK MANAGEMENT INTO THE 21st CENTURY: Techniques, Standards, Technologies, and Applications
Salah Aidarous and Thomas Plevyak
1994   Hardcover   448 pp   IEEE Order No. PC3624   ISBN 0-7803-1013-6

ENGINEERING TOMORROW: Today's Technology Experts Envision the Next Century
Edited by Janie M. Fouke, written by Trudy E. Bell and Dave Dooling
2000   Hardcover   288 pp   IEEE Order No. PC5802   ISBN 0-7803-5361-7

THE DISAPPEARANCE OF TELECOMMUNICATIONS

Roberto Saracco
CSELT—Telecom Italia Group

Jeffrey R. Harrow
Compaq's Technology and Corporate Development Organization

Robert Weihmayer
GTE Internetworking
Cambridge, MA

IEEE Communications Society, *Sponsor*

The Institute of Electrical and Electronics Engineers, Inc., New York

This book and other books may be purchased at a discount
from the publisher when ordered in bulk quantities. Contact:

IEEE Press Marketing
Attn: Special Sales
445 Hoes Lane, P.O. Box 1331
Piscataway, NJ 08855-1331
Fax: +1 732 981 9334

For more information about IEEE Press products, visit the
IEEE Press Home Page: http://www.ieee.org/press

Printed in the United States of America

10   9   8   7   6   5   4   3   2   1

ISBN 0-7803-5387-0
IEEE Order No. PP5404

Library of Congress Cataloging-in-Publication Data

Saracco, R. (Roberto)
   [Scomparsa delle telecomunicazioni. English]
   The disappearance of telecommunications / Roberto Saracco, Jeffrey R. Harrow, Robert Weihmayer.
     p.   cm.
   Includes bibliographical references and index.
   ISBN 0-7803-5387-0
     1. Telecommunication—Popular works.   2. Information society—Popular
works.   3. Telematics—Popular works.   I. Harrow, Jeffrey R., 1949–   II.
Weihmayer, Robert, 1953–   III. Title.

HE7631.S2713   1999
384—dc21   99-052048
                                  CIP

Dedications

I dedicate this book to my colleagues at CSELT, who provided continuous stimulation and depth by sharing their research activities and results, and to Cesare Mossotto, CSELT's director, who understands the thin line separating work and pleasure and who let me extend my field of interest beyond telecommunications.

I dedicated previous books to my wife and children for bearing with me. They keep doing so in spite of my promising "this is going to be the last." Thanks to them and to Luisa Rossari, my invaluable secretary, who keeps helping and saving me time to dedicate to my family.

Roberto Saracco

I dedicate this book to Fran, for her love, encouragement, and support and for putting up with the long hours. Also, to my Uncle Ralph, for sparking my passion for technology at an early age. And to Wendy, for giving me the continuing opportunity to keep my fascination with technology—and my desire to share it with others—alive.

Jeff Harrow

I dedicate this work to my immediate family: my wife, Sandra, and my two young daughters, computer experts in the making, Erika and Melissa, for their infinite patience. I hope that many teachable moments will come out of this book.

Robert Weihmayer

Contents

Foreword

When I mail a postcard from a distant land, I consider the pure magic of how it arrives at its destination. Imagine the handling involved as it takes various flights, joins other correspondence, and ultimately is sorted to arrive at one specific mailbox at a more or less random place on the planet.

In contrast, when I send an Email, it is truly simple to comprehend how those weightless and sizeless bits can be routed at the speed of light, although far less magical.

Our great-grandchildren will feel this even more dramatically, as a postcard will cost $50 to mail and telecommunications will be free of charge. What does this mean?

Air is free. At least it is free if you are not below water and are willing to accept the outdoor climate. However, as you add value to air, it starts to cost something, whether by removing humidity, adding scents, changing temperature, removing particles, or the like. Also, we really don't notice air until it is missing, very dirty, too hot, and so on.

Likewise with telecommunications. This book is about the vanishing frontier of telecommunications, a new epoch when computers will become unnoticed. People who deliver bits will prosper; those who merely hold them will be out of business.

Why is this happening now? Simple. Because communication is digital and because it is moving from circuits to packets, which is an architecture that makes metering a mockery and distributes telecommunications infrastructure both more efficiently and asynchronously. Capacity is shared in ways previously impossible.

People are also changing the future of the future. A massive amount of the world is using the Internet knowingly and many more unknowingly. Almost every forecast of future use is low and the official forecasts are mostly off the mark. If you hear an estimate, you can safely double it.

Aside from the raw numbers, including the scale of Ecommerce, another event is happening that our grandparents would not understand. The digital world has taught us something about change itself: what we thought to be unchangeable suddenly is. Basic assumptions about how we live our lives, learn new ideas, or conduct business are all subject to change. Creating a five-year plan is a fine exercise, but the contents are invariably a bad joke. We know that.

Those who think otherwise should read on. Those who agree, should read on anyway.

Nicholas Negroponte

Preface

This book was first published in Italy. The authors[1] were all from CSELT, the research branch of Telecom Italia, and that edition focused on an Italian audience. Because the book generated so much interest in Italy, I discussed the possibility of an American version with some friends at the IEEE Communications Society. The idea was received with enthusiasm.

From the very beginning, it was clear that this would not be just simply a matter of translating the text from one language to another; rather, it would take into account the different cultures, and the significant technological and market changes that have occurred in the past two years. For instance, consider the culture issue. Italy is an advanced European country in terms of telecommunications infrastructures. Just witness the astounding explosion of cellular phone activity: by July 1999, Italy had more cellular telephones than fixed ones! On the other hand, Italy does lag behind in terms of the emerging Knowledge Age, probably due to a culture factor. Consider this, regardless of the humorous Italian gestures depicted in movies, Italians do indeed prefer to communicate by talking to each other face-to-face, complete with waving hands and smiling :-) or frowning :-(faces.[2] That bit of culture has had an impact on how Italians have embraced technology.

Many years ago, I was part of the group that established the first electronic telephone office in Italy. As it came on-line, we went to a village in the countryside to see it in action. In those days of mechanical step-by-step telephone offices, it was always easy to monitor telephone traffic—the noisier the relays in the building, the more people were dialing! Modern computerized exchanges are usually silent, except for the muffled sounds of cooling fans, and we could see only the flickering of little red and green lights to show activity. As we stared at the pattern of those flickering lights, we realized that something unusual was happening. We expected to see a typical conversation lasting for several minutes, but instead we noticed a preponderance of unusually brief calls. Wondering if there might

1 Roberto Saracco, Michela Billotti, and Margherita Penza

2 These two odd constructs :-) and :-(are called *emoticons*, or *emotional icons*, and are good examples of how the Internet is changing the way people communicate. Because it is difficult to clearly express emotion through text, these special shorthand markers evolved to facilitate communications on the Internet. But in true Internet fashion, we now see emoticons in books, movies, and even billboards. See www.geocities.com/MotorCity/Pit/4824/smileys.html for a list of common emoticons.

be a problem with the connections, we eavesdropped on a call and over-heard this fragment: *"Giovanni, it's me, Mario. Come down to the street, because I need to talk to you."* The phone might serve in getting an Italian's attention, but it seemed the actual conversation occurred face-to-face. Although this happened many years ago, it demonstrates the discomfort that some Italians still feel with entrusting their conversations to the wires![3]

Thus, it may be more a cultural discomfort factor than a shortfall on the technology front, which plays a role in Italy's slow acceptance of the Ecommerce and Ebusiness practices that have taken North America by storm.

These cultural issues compelled me to seek a partnership with people who lived and worked in an American cultural context to make the English version of this book relevant to that part of the world. If I could find people who shared my beliefs in the "disappearance of telecommunications," I could seek assistance in adapting and updating the book.

I approached Robert Weihmayer with the idea of this collaboration. We have known each other for the past ten years from various IEEE workshops and conferences, and shared a common interest in network operations and management. He was excited by the proposal and promptly accepted. Jeff Harrow's thoughts in the weekly technology journal, the *Rapidly Changing Face of Computing,*[4] have been a constant inspiration to me. He follows the innovations and trends of computing and speculates on their direction. Even though we'd never met, in a sense he was a virtual acquaintance made possible by the Internet—a proof that global communities do indeed exist. He also accepted my proposal and thus we formed a team.

The three of us bring widely differing backgrounds, opinions, and insights to this translated and updated edition, and we believe that our discussions and brainstorming bring you more than the sum of the parts. By the way, these are expressly the views of the authors and not necessarily those of their employers.

I'd like to thank Jeff and Robert for their valuable contribution to this revision. Most of the notes appearing throughout this book have been updated to reflect the many innovations that are proof of our rapidly changing world.

---

3 But this does not, apparently, apply to phone conversations without wires! Italians fell in love with their cellular phones ("telefonino"), with recent estimates indicating an 80% penetration of cellular phones for those aged 14 and over by 2003.

4 See www.compaq.com/rcfoc

Finally, a grateful thanks to my original coauthors who worked with me to prepare the Italian edition; they were as invaluable in their contribution to the Italian edition as Jeff and Robert have been in revising this translation for English-speaking cultures.

Roberto Saracco

The Disappearance of Telecommunications

We tried to come up with a provocative title, something that would tickle your curiosity. We also wanted a title that, no matter how strange it seemed at first glance, would spark a picture of what is to come. "The Disappearance of Telecommunications" fits both bills.

Paradoxically, we, the authors, are actively involved in the world of telecommunications and computing, so the last thing we'd want is for telecommunications to disappear! Nevertheless, as perceived by a large share of the population, that is exactly what we forsee.

INTO THE LANDSCAPE   The easiest way to appreciate this is by example. Think for a moment about the roads you travel every day to school, to the office, or to the mall. How many times do you find yourself thinking, "Hey! There's a road!"? Rarely. It's just human nature that the things we use as a part of our daily routine simply "disappear" from our perception.

IT'S THE COMMON THINGS AROUND US THAT WE TAKE FOR GRANTED THE MOST!   The coming together of new telecommunications and computing technologies into new products and services, which seamlessly integrate into how we work and live our lives, seem destined to become so familiar and so much a part of our environment that they too will fade from our perception.

Of course things are not always so transparent—when we're driving down the road and get caught in a bumper-to-bumper traffic jam, we do indeed start (unflatteringly) thinking about the road—how we wish it were wider or that there were more alternate routes. Similarly, it's only when our "expected" telecommunications services break down (an unusual event)[1] that we really think of them at all.

Wireless phones are expanding at an incredible rate. Worldwide, there were about 880 million fixed phones and 380 million cellular phones in June 1999 (Salomon Brothers). Those figures are expected to increase to 1.2 billion fixed and 600 million cellular phones by the year 2005, and by 2010, the total number of phones is expected to be 2.4 billion, half of them fixed and half cellular! Wired phone lines are actually expected to decrease in some developed countries!

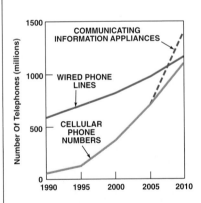

1 Our traditional telecommunications infrastructures are, truly, "built for the ages," such as telephone exchanges, which must guarantee no more than one hour of downtime over 40 years! To get a better idea of just how reliable the phone service is, consider this. If your computer crashes, you may be upset but, sadly, certainly not surprised, but you'd be shocked if you didn't hear a dial tone the next time you picked up the phone!

Today in developed countries, the telephone is part of everyday life. We rarely notice it; although new services, such as videoconferencing, still cause most of us to raise an eyebrow. Yet that too will pass; the day will come when we will think no more of a videoconference than we do of a phone call, at which point videoconferencing will also disappear from our notice. Just think back to when a cellular phone was an ultra-expensive business tool—people took notice when they saw one in use. Today, cell phone users dot the landscape like trees.[2]

There is another element, beyond a disruption of service, that causes us to notice the otherwise hidden bits of infrastructure around us: cost. We notice the road when we have to stop for a tollbooth. At home (if not at the office we usually do think about the cost of a long distance call, because a business call is now far less expensive, and more timely, than a traditional business letter). Cost affects how we use things, but the future will hold some very interesting surprises as far as the cost of telecommunications is concerned![3]

So telecommunications is becoming more pervasive and less expensive, and its services are insinuating themselves into more of the elements of our business and personal lives. As this progresses, like the asphalt roads before it, telecommunications will sink below our notice—it will disappear. Strangely, it's *just* this disappearance that will mark the real beginning of what we refer to as the Knowledge Age.

Let's look at an example of how one business altered its model through embracing telecommunications. A florist[4] in the United States grew beyond its physical reach by offering to send flowers around the world, using the phone to see that the order would be delivered by a florist close to the recipient. They didn't actually deliver their own flowers (atoms), but caused an equivalent to be delivered by extending the order over the phone (bits).

Their next step was to offer a customized service of sending flowers based on a customer-defined schedule, so the customer no longer had to worry about remembering friends' and relatives' birthdays. Here they leveraged the information they had collected from their customers (bits) and used it to help (and so retain) those customers' business.

2 Like trees, the proliferation of cellular phones tends to obscure the landscape—many restaurants, movie theaters, schools, and other public places are beginning to make pocket phones feel unwelcome.

3 For example, as we write this in mid-1999, people in the United Kingdom are, for the first time, experiencing Internet access *without* having to pay the per-minute telephone charges for local calls that are common in many countries. This altering of the cost of telecommunications has had a very real effect—10,000 new U.K. Internet citizens *each day*— news.bbc.co.uk/hi/english/sci/ tech/newsid_327000/327010.stm and the United States is now in a price war, driving domestic long distance calls to 5 cents/minute.

4 1-800-FLOWERS

Of course if they could use this model for flowers why not for other products such as chocolates, still focusing on their customers' real need, to send someone a gift. Also, why not leverage the telecommunications network even further, by delivering a gift that includes a toll-free phone number, which plays a personal message recorded by the sender?[5]

This real-life example, which can easily be extended to other sectors of our economy, shows how telecommunications is broadening beyond the phone to become only one "invisible" feature of a very broad spectrum of service packages.

MORE CHANGES AFOOT   Today, although telecommunications is beginning to disappear, it's still far from invisible. For example, if you're using the Internet at home and you want to look up a phone number in on-line white pages, you probably have to first dial your ISP. Thus, it's probably faster to look it up in the paper directory. However, if you have one of the emerging "always on" Internet connections, such as a cable-modem or an ADSL connection, then looking up that number becomes a simple and fast operation, because your Web browser is "always live," and you'll probably never reach for the paper directory again. Here is one example of how the use of a telecommunications service (when it is instantly available and the cost is already absorbed; in this case, the monthly ISP bill).

THIS ROAD AHEAD   Throughout this book, we will continue to explore a wide variety of services that we believe could become a part of our everyday lives within the next ten or so years; some, or the technology that could lead to them, are already available in the laboratory. Although this book isn't about predictions, some will come true and others won't, depending on the ebb and flow of ideas from ingenious startup companies and established research firms and how the marketplace embraces those ideas. This book, instead, is about opening a window for you into tomorrow's possibilities. Perhaps by looking through this window, you will be the one to make them happen!

As you read this book, don't feel bound by the order of the chapters. Whereas we do endeavor to tell a story, feel free to explore hither and yon for the areas of most interest to you.

At the end of 1997, "data" was still a small percentage of the world's total telecommunications traffic, but in 1999 it is enjoying tremendous growth. WorldCom-MCI estimates that data traffic represented 1% of the total traffic handled on the worldwide telephone network at the end of 1996, but it's expected to hit 50% by 2001! Indeed, in 1999, US data traffic already accounts for 40% of the total telephone network traffic. By 2004, most voice traffic in the world will simply be processed as data (which, in fact, it is). The Internet, which by the end of June 1999 included 179 million "citizens," will, of course, be responsible for a large part of this data transmission growth (www.nua.net/surveys/how_many_online/index.html).

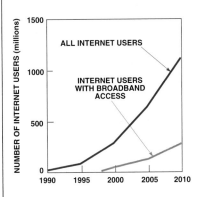

5 And watch out—as cell phones become even less expensive, we can imagine that the telephone number will be replaced by a "disposable" cellular phone that will automatically call the person who sent the flowers!

Television remains one of the most widespread means of communications, and it will become increasingly interactive over the next few years, thanks to its convergence with the Internet.

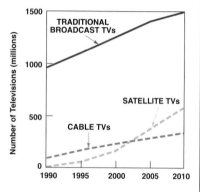

This book mirrors the authors' optimistic views of how telecommunications and computing will be affecting the way we work, live, and play. However, we are fully aware that there are many sides to every issue and that all sides, positive, negative, optimistic, and pessimistic, must be explored for a full understanding.

THIS 'TELEPHONE' HAS TOO MANY SHORTCOMINGS TO BE SERIOUSLY CONSIDERED AS A MEANS OF COMMUNICATION. THE DEVICE IS INHERENTLY OF NO VALUE TO US.— WESTERN UNION INTERNAL MEMO, 1876

We also recognize that some of the concepts and ideas presented here might seem like science fiction, perhaps even downright absurd, so we're offering you many references, most on the World Wide Web, where you can glean more information. Although all these references were validated at the time of printing, due to the nature of the Web, some might no longer be accessible when you read this. If that happens, a little work with a good search engine, such as Alta Vista,[6] should help, and we plan to keep references updated in an electronic version of this book available on the Web and we plan to keep references updated on our book's website.[7]

THE WIRELESS MUSIC BOX HAS NO IMAGINABLE COMMERCIAL VALUE. WHO WOULD PAY FOR A MESSAGE SENT TO NOBODY IN PARTICULAR?—DAVID SARNOFF'S ASSOCIATES RESPONDING TO HIS URGINGS FOR INVESTMENT IN THE RADIO IN THE 1920S

We value your feedback and questions. Please visit our discussion group on the Web[8] and explore, argue with each other, teach, and learn.

Of course you can also send your comments to us via Email at telecom@cselt.it.

---

6 Alta Vista—www.altavista.com
7 See www.ieee.org/press/authors/ saracco
8 See www.ieee.org/press/authors/ saracco

Reflections: Economics of the Knowledge Age

To get a sense of how deeply the Knowledge Age is taking root, we need only to consider its impact on the economic front. Three areas, in particular, tell us a great deal: the rapid expansion of the use of computers, the penetration of the Internet, and the vast growth in the capacity of our global data networks.[1]

Nowhere else is the impact of these economic factors more apparent than in the United States. By March 1999 the numbers were already staggering: a 50% household penetration rate for personal computers, a 26% penetration rate of the Internet, and approximately 350 gigabits per second of raw network capacity available! These numbers indicate why the United States has emerged as an influential driving force behind the Knowledge Age. In addition to this, innovative U.S. businesses such as Microsoft, AOL, Intel, Amazon.com, and others are creating new and forceful market trends that uniquely define and dominate the economy of the Knowledge Age.

The penetration of personal computers and application development capacity—very important elements for the diffusion of information in the Knowledge Age—are much higher in the United States than in Europe and Japan.

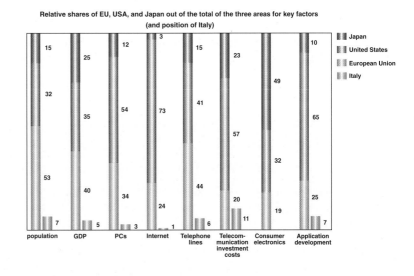

Relative shares of EU, USA, and Japan out of the total of the three areas for key factors (and position of Italy)

---

1 By about 2003, it is estimated that over 40% of the devices used to access the Internet will *not* be personal computers; they'll be augmented with interactive televisions, personal digital assistants, and other focused information appliances. By 2010, it is being predicted that 10 "information appliances" will be sold for every personal computer. It's not that PCs will be going away, but there will be many more devices joining them in reaching out and touching the Information Highway!

DRUMBEAT OF THE KNOWLEDGE AGE The coming together of computers, communications, content, and consumer electronics, or "convergence," which is driving the Internet, is really the sculptor of the Knowledge Age. Let's explore some of the economic and, by extension, social aspects of this process.

This convergence of computing, communications, content, and consumer electronics resonates in all walks of life: from education to health care, social services to professional services, and from telecommuting to virtual markets for goods and services. In all these areas and more, human productivity stands to increase dramatically, from instant access to the knowledge of a huge number of people and ubiquitous access to information from every aspect of our society.

This advent of the Knowledge Age has significant implications to our economic system. The transformation of the way people work and the way business is done will directly affect the economy. As demonstrated by Amazon.com becoming the third largest bookseller in the world almost overnight, it is already changing how and where consumers buy and the balance of power in entire industries.

Clearly, not every aspect of the economy will be affected in this same way and at this same rate, but nowhere are these changes more visible than in those industries where information gains "value" from being transformed, or through simply being made available online.[2]

For such businesses, the opportunities for creating wealth will be increasingly knowledge based. Those wishing the best competitive advantage will strive to leverage the most advanced technology and most creative human resources available for adding value to their information in ways their competitors cannot. Let's consider a few examples.

MY HOME IS MY CASTLE Telecommuting will have a drastic influence on the nature of the workforce available to businesses, and it will lead to a restructing of how business is "done." It may even give rise to a new class of worker—the "telecommuting knowledge professional."

While not tied to a specific title or job description, this type of worker is beginning to restructure their relationship to their work, and to their employers. Quality of life is an increasingly important factor to people, and traditional ideas and constraints about how and when work is performed will become a thing of the past. For many (not all) jobs, time schedules, offices located in specific areas, congested cities, and that dreaded rush-hour traffic might become a bad memory! And these are just a few of the Knowledge Age fringe benefits.

---

2 For example, Amazon.com. See www.amazon.com/

Another consequence is that people will be valued and used in very different ways by Knowledge Age businesses, and more people, those living far from traditional commerce centers, will have the opportunity to become prime contributors. This may lead to some interesting fallout. As people begin to live where they want to, rather than having to crowd into and around cities, could we end up with semideserted urban office parks, lots of empty parking spaces, crumbling city real estate values, and lowering tax bases? On the other hand, traditional "vacation communities" might become year-round havens, spreading wealth throughout the land! Satellite Kinko's "remote office centers" might provide those high end or bulk business services that automated home offices cannot provide.

All of this could be the result of ubiquitous, inexpensive, high-speed, reliable telecommunications.

SPEED DOES MATTER The business skills to train and manage this new type of Knowledge Age worker, and to manage new technologies, will become a decisive factor that will drive economic development in both industrialized and developing countries. In fact, it may be those "developing countries" that are able to reap the *most* benefits from the waning importance of distance; they may be able to leapfrog the growth of today's industrialized nations to achieve a similar lifestyle in far less time![3]

Let's look at a specific example. Historically, it took 58 years to double the average per capita income in the leading industrialized nation, the United Kingdom. The United States achieved the same result in 47 years. Germany and Japan took 43 and 34 years, respectively, but since 1966, it has taken Korea only 11 years, Chile, 10, and China, 9 years for the same per capita income growth. Clearly these data are open to interpretation, but there's no doubt that the rate of change is steadily increasing. This implies greater opportunities for those countries that can keep pace—yet even bigger problems for those that are left behind, because already developed countries will pull even farther ahead of those still waiting to begin their transformation.

Those countries that develop Knowledge Age skills, products, and services will enter the global economy that much faster, but to do so they must develop the ability to effectively use new technologies and new organizational skills. Launching a dramatically improved economy that trades on a par in the global marketplace could be worth the effort.

CUT THE MIDDLE PERSON Let's continue with a broad look at how the Knowledge Age is likely to affect business.

---

3 For example, Bangalore, India has seen a tremendous increase in job opportunities and wealth by becoming a huge software factory to the world.

It seems inevitable that Convergence will make people and information-based resources more productive, and rather than displacing human workers, we think this knowledge-based environment will demand more humans, with greater skills—just not always the *same* skills that were valued in the industrial Age.

Some jobs will go away. Consider the "intermediaries" that have typically created layers between suppliers and consumers. For example, there used to be many travel agencies in the United States that did little more than "write tickets," but this was a valuable service because nowhere else could a traveler gain access to all the airlines' schedules and fares in one place. Then, enter the Internet with sites like Expedia and Travelocity[4]; they provide all this information and more with a click 24-hours per day! Travelers are flocking in droves to this new more convenient way of buying tickets, and this is "disintermediating," or displacing, those travel agents that didn't offer value above and beyond simply writing tickets.

On the other hand, while the Industrial Age travel agent *is* being "disintermediated," it's interesting to note that these new Knowledge Age travel services are *themselves* "intermediaries"—but of a different sort! So, it's not that intermediaries will go away, only that they are being transformed to add value in this new environment. It's a shift from the physical to the virtual.

As Convergence continues to introduce new efficiencies, we will see a continuing reduction in transaction costs,[5] broader (global) markets, and vast new opportunities for business! This transformation will require the development of new retail techniques, easier ways for customers to do business, and new ways for providing satisfying after-sale customer services.

RESPECT FOR THE LAW   As with the Industrial Age, the legal infrastructure of the Knowledge Age will be a fundamental enabler of its success. New ways of doing business for producers, consumers, and property owners (including intellectual and information-based property owners) will require new rules of engagement and new forms of negotiations; new legal frameworks will have to evolve to provide effective checks and balances.

There are many factors. Can you prove that an electronic document hasn't been altered or that it was actually sent by one particular person, and received by another? How do you audit electronic commerce to the satisfaction of courts and shareholders? Are electronic commerce processes secure and reliable? To answer these questions and many more, today's Industrial Age policies must evolve into a stable and robust regulatory framework to guarantee that the Knowledge Age economy will work predictably for all players.

---

4 www.expedia.msn.com/ and www.travelocity.com

5 A bank transaction at an ATM costs the bank just a few cents, whereas a similar transaction with a human teller costs many dollars.

Indeed, the basic relationship between governments, public administrators, and citizens is already beginning to change. Utah's governor has signed into law the "Digital State Act," which requires every state agency provide their services online within three years! This includes online ways to renew your driver's license, obtain a hunting license, register your car, and even file court documents! Countries such as Brazil, New Zealand, Holland, Singapore, Chile, and Portugal already allow their citizens and businesses to directly file their taxes online!

The residential market focuses increasingly on content, whereas the business market tends to favor applications. For both, the percentage of hardware is expected to decrease to a great extent due to the reduction in the costs of equipment and infrastructures.

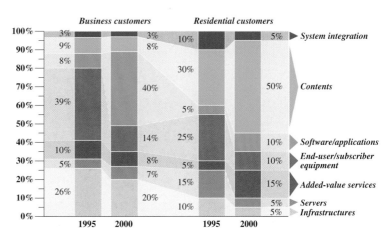

BOOZ-ALLEN & HAMILTON analysis, European industrial analysis

The Knowledge Age is already very tangibly changing the way that people and governments interact, and this is just the beginning.

THE KNOWLEDGE AGE WORKER

Not only is the Knowledge Age changing how governments and people interact, but it's also changing the dance between employers and employees as Convergence unfetters how and where people can best work. Telecommuting is allowing people to work where the job can best be done (perhaps on a customer's site, in a car, or at home). Old ways of supervising and measuring employees no longer apply; new ways, focusing on the quality and quantity of work delivered rather than on which hours someone sits at a desk under a manager's watchful eye, must evolve, and the creativity and innovation that people can bring to solving a problem must be cherished and nurtured.

"A company that wants to remain ahead has to institutionalise constant innovation!"—Peter Drucker

1   The Knowledge Age

Information is ethereal. It's intangible. In many ways, it's highly subjective—information for one might be noise for another. Information might not be perceived to be a valuable commodity by some, but today, in some ways, information is being valued far more than tangible goods by others! Consider how the market valuation of some "Internet companies," which don't actually "make anything," far exceeds the value of many successful and long established Industrial Age manufacturing companies! They turn raw information into knowledge and are rewarded for doing so. Welcome to the Knowledge Age.

How did we get here? Consider that during most of human history we lived in the Agrarian Age, where a vast majority of people toiled in the fields to feed themselves and a few others. Change happened so slowly, it seemed invisible; many generations did things the same way as their grandparents.[1]

Then, about 150 years ago, factories, and initially steam and water power and later electricity and mass production, changed the rules. People left their farms and congregated in bustling new towns built on river banks where water could drive the factories' machines. Industry evolved, providing a vast new array of affordable goods, and society changed, because the social rules that worked so well for families spread out on farms just couldn't apply in the very different world of the mill towns. Parents were at a loss in understanding all the new ideas and customs, the "change" that their kids brought home, because the parents really had done things the same way as their grandparents, and that had worked just fine in their environment. They just weren't prepared to understand how much, and how quickly, their kids were changing.

Similarly, we're now entering the Knowledge Age. Instead of water, steam, and electric power causing this new age, *this* revolution is about the coming together of communications and computing to enhance the value of "information." Have no doubt, this Knowledge Age revolution is

1 We'd like to acknowledge the insights and ideas of Rich Lang of Compaq for having started us thinking about many of the ideas in this chapter.

having every bit as dramatic an effect on us—on how we work, live, and play—as the Industrial Age had on our grandparents.

How might the production of goods and services change, say, over the next 20 years? An MIT study[2] suggests the possibilities of two rather different paths.

SMALL IS BEAUTIFUL

Scenario 1: Toward a Society of Entrepreneurs

Easy access to inexpensive and pervasive computing and communications will enable everyone, even small groups and individuals, to develop businesses that once required the resources of large corporations. Society will be based on micro-enterprises. The ability to differentiate, based on individual creativity, will become increasingly necessary to ensure success.

It will also be possible to offer more complex products and services by reaching out through the Internet, or what it becomes, to leverage whatever competencies are needed for a given project. One contemporary example of this might be the Open Software movement, where such products as the Linux operating system and Netscape's browser are created by scores of ad hoc developers, each doing what he or she wishes and folding their work back into the core product.[3] Traditional software developers, even very large ones, are paying careful attention.

Of course this can't work for every or even many products today. Only in the ethereal world of information products can people work without factories or other more substantial ways of turning their work into tangible reality—but there are certainly a growing number of information-based products representing an increasing part of the economy.

Desktop Manufacturing?

Actually, that previous statement is no longer entirely true. Although not yet available as a $100 peripheral at your local computer store, techniques such as "stereolithography" do make it possible for a 3D computer model to be rendered into a tangible solid without using any traditional machining tools![4] One technique uses a tank filled with a light-sensitive liquid epoxy. A special laser "writes" the 3D image of the object to be created from the "inside out" in the tank, and where the laser touches, the liquid solidifies. The end result is a part made to within a couple of thousandths

---

2 The ideas presented come from R. J. Laubacher and T. Malone studies at MIT on "Inventing the Organizations of the Twenty-First Century." In a first paper in 1997, Working paper 21C WP001, they took a neutral position with respect to the two scenarios, but in a more recent paper in the Harvard Business Review, September–October 1998, they clearly point to the Society of Entrepreneurs, calling it "the dawn of the e-lance economy" (e-lance is short for electronic freelance).

3 "The Cathedral and the Bazaar"— www.tuxedo.org/~esr/writings/ cathedral-bazaar/cathedral- bazaar.html

4 For example, one company providing such services right over the Internet is Scicon Technologies—www. scicontech.com/ Email them your 3D model and your actual part comes back a few days later via FedEx!

of an inch of the computer model and in a wide range of textures. If this equipment gets less expensive (and remember laser and ink jet printers were once very expensive), can you say "desktop manufacturing?"

Speaking of intangible information-based products, they're a lot easier to steal and copy than physical products—for such an information-based economy to flourish, the builders of these information products have to feel confident that intellectual property laws will respect the value of their work. (Hence the brouhaha from the traditional music distributors over MP3, a file format that allows people to send CD-quality music over the Internet with no muss or fuss—and sometimes without the legal right to do so.)[5] Indeed, as we write this in mid-1999, "MP3" has become as popular as "sex" as an Internet search term![6]

BIG IS BETTER

Scenario 2: Global Mega-Corporations

In this possible future scenario, complex, costly, centralized, and constantly evolving production facilities will be required to produce products and services, and these huge infrastructures confer a competitive edge to their companies. Only large companies will have the power and wealth to create and maintain these infrastructures; indeed, groups of such companies will leverage each others' competencies to provide even greater competitive advantages (in the manner of the Japanese keiretsu), making the price of entry into a market so high that only the mighty can play. Society will be dependent on the mega-corporations, to the point where they replace individual states and countries as they are known today!

WHICH WAY WILL IT GO?
That, of course, is up to all of us, and both scenarios may coexist, but it is interesting to note that both scenarios require very sophisticated communications and computing infrastructures. In the first scenario, the rapid and free-flowing interaction between micro-enterprises, required to support such ad hoc businesses, would not be possible without Knowledge Age support. Similarly in the second scenario, it would be impossible for the mega-corporations to even exist without global high-speed communications and computing.

The growth of a pervasive high-speed communications infrastructure may change the rules in other interesting ways—consider that the geographic population clumping of the Industrial Age may no longer be nec-

5 "Rapidly Changing Face of Computing" April 19, 1999—www.compaq.com/rcfoc/19990419.html#Online_Music_Feel_The

6 "How Digital Music Could Change Your Life," ZDNet News, April 30, 1999—www.zdii.com/industry_list.asp?mode=news&doc_id=ZD2250087

essary and that could have far-reaching effects on cities, states, and nations. In fact, we're already beginning to see the first signs.

For a growing number of jobs, the time-honored tradition of "going into the office" is no longer required. Even for people still working for large corporations, the advent of notebook computers, fast dial-up modems, and wireless connections enables them to receive many traditional and enhanced office support services from where they can best do their jobs. They may work from a desk in their customer's facility, from their car while moving between customers, or from their home, but their corporate services, and an Internet full of public information, is but a click away. Who benefits? Many employees find that this flexibility improves their quality of life, and businesses directly save on real estate and other costs associated with traditional office space, as well as on employee travel time.

Of course this may yield some interesting side effects. For example, a business may be able to hire talent anywhere on the planet, rather than just those people who happen to reside (or are willing to move) near an office building. Does this mean that a talented individual who happens to reside in a developing country with historically low salaries (compared to developed nations) will now demand equivalent pay, globalizing salaries?

How about where people live—today we clump together in and around cities because that's where we've had to be to work, and often, people can't wait until they can "get away from it all" on vacation. Tomorrow though, might people be able to live at the beach, in the mountains, or on an island while still participating in the global economy? And if they do, would they then spend their vacations visiting the "big cities" that have then become the novelty? Imagine the impact on tax bases, real estate values, and more. Will hotel "meeting facilities" become passe if virtual meetings become commonplace? Might hospital outpatient clinics shrink if patients can receive ongoing treatments, such as kidney dialysis, at home with the equipment monitored remotely? (This is already beginning in Italy.)[7]

How about home automation, which today is mostly in the domain of the "enthusiast"—it could become far more common. The evolution of technologies such as X-10, CEBus, and HomeAPI may soon have you expecting to be paged when someone rings your doorbell. Your wristwatch "pager" would show you a video of who is at the door, you could chat with her, and even remotely open the door so she could leave you a package.

---

7 San Giovanni Hospital in Turin and Policlinico Gemelli in Rome, Italy.

The Knowledge Age has the potential to bring many new services into our lives, and do it in ways that reduce the effort we have to expend. For example, today when you order flowers over the phone, you don't know or care which actual florist satisfies the order at the other end. Similarly, today you can order a pizza on the Web and, based on your address, the company's Web application will forward the order to the closest pizzeria.[8]

BACK THE OTHER WAY

The Industrial Age moved the focus from the individual craftsman to the assembly line. Individual wants and desires had to take second place to the choices offered by big business (Henry Ford was willing to sell you a Model T in any color you'd like—as long as you liked black), but the Knowledge Age is reversing that process.

For example, Levi's has been experimenting with a process in which you go to a store and get carefully measured. Your measurements and your choice of a jeans style from a book are entered into their computer system, and a custom–fitted pair of jeans ends up on your doorstep a couple of weeks later. Another company offers similar options for custom-made shoes.[9] Also, Motorola never prebuilds certain lines of consumer pagers—your order goes directly to the manufacturing line, which turns out the pager you want in your color, playing your choice of tunes when it gets paged, and so on.

THE DISAPPEARANCE OF THE "AVERAGE" CUSTOMER

This trend, sometimes called "mass customization," could go much farther. Suppose you went to a "scanning booth" at the mall and had your entire body scanned in 3D.[10] This model of your body would be placed on the Internet (under your lock and key), and you could "bring" it with you as you visited various clothing manufacturers' Web sites. See a shirt you'd like? With a click you can see it right on your body, draping just as the real garment would. If it "almost" fit, you could order one made to your exact specifications, because your 3D model would provide your measurements directly to the manufacturer.

I'LL MAKE THIS FILM!

Movies may get even more interesting—some producers are experimenting with putting a keypad on the seat in front of you in the theater—at various points in the film the audience votes on what they'd like to happen, and the story then unfolds in that direction. These are crude right now, but later films will get

---

8 Pizza Hut is providing "order on-line service" on an experimental basis in Topeka, Kansas. www.pizzahut.com
9 Digitoe—www.digitoe.com/digitoe/index.html
10 See www.compaq.com/rcfoc/19990927.html#_Toc462668375

11 Work on such vision systems is currently being done at MIT's Media Lab www.media.mit.edu/affect/, and such interactive movies may cause interesting social changes as well—clubs of like-minded people may band together to go to the show at the same time to influence its direction based on their common interests, such as the "Fantasy Club," or another club that chooses to minimize violence.

12 See www.compaq.com/rcfoc/ 19990906.html#Toc_460897397

more sophisticated in their alternate paths, and computer vision systems may watch the audience and react to their emotions and interest/boredom levels without requiring keypads.[11] But imagine the incentive to the movie studios, hoping to get you back to see (and pay for) a favorite film many times to explore all its possible twists and turns!

You see, in many ways, a communications- and computing-driven Knowledge Age may be transforming us from spectator into actor, from someone buying "off the rack," to someone demanding custom tailoring. This mass customization may create vast opportunities for cottage industries to meet the demands.

THE DARK SIDE

Of course there is a potential dark side to this Knowledge Age force, where most everything we do, every transaction we make, is digitized, transmitted, and stored in a way that may well identify us. Credit cards leave an obvious trail, but consider the "loyalty cards" we use to get extra discounts at grocery and drug stores (they give the store a detailed picture of our buying habits—and don't forget every phone call we make).[12]

New definitions (and safeguards) for our privacy will be tremendously important. New ethics will be required to set the boundaries between what is permissible and what violates our personal space. These aren't new considerations by any means, but with so much personally identifiable information available so easily to so many, they will take on a far more increased importance.

Organization and Business Reengineering

DOING IT MY WAY For most of recorded history, the production of goods and services was very personal: a craftsperson creating each item individually with the skills taught by years of apprenticeship, for instance. Individuals strove to improve the quality and quantity of their products everywhere from collecting the materials to putting on the final finish—from beginning to end. Take farmers, for example, who had to know all the ins and outs of their business, from plowing to seeding and from weeding to harvesting, and then, finally, going to market.

The advent of the Industrial Age made it necessary for humans and machines to interact in a far more complex and intertwined dance, with individuals focusing on improving an ever-smaller subset of the end-to-end process; thus was born *specialization.*[1]

DOING IT THEIR WAY The rise of ideas such as Taylorism[2] came about as people strove to make the new factories and mass production systems work better. In this context, optimizing these various specialties became the key to business success; essentially, every element of a job was intensely studied to get rid of every wasted motion and wasted moment, to get the most work out of each individual—to keep all these human cogs in the factory machine working in sync. Everything required for a particular task would be identified, and processes put in place to assure that they were available where and when needed. Each movement of a job would be choreographed to assure no time was wasted. A job's result would have to be made available for the next step of the process when and where it was needed.[3] But people don't make good machines; they're not generally well cut out to work in this exceedingly boring machine-like manner, which has been likened to a corporate battle cry of *"I don't understand, but I'll do what I'm told."* Indeed, such conditions result in frustrated workers and managers alike.

---

1 Today our society and the processes that make it work are so specialized that it has become infeasible for most people to "go it alone." Imagine having to hunt for food, chop trees to build a house, grow cotton or shear sheep for thread to weave fabric, or develop your own medicines. Even making the most commonplace item, such as a glass, is beyond most of our skills. In a way, we've become as specialized as ants and other communal insects!

2 The "Time and Motion" studies were introduced by Frederick Taylor at the beginning of the twentieth century to measure and improve the efficiency of industrial workers in the United States.

3 This was humorously depicted in the 1936 Chaplin film "Modern Times."

THE BEST OF BOTH? Today, work processes are often restructured to make the most of human strengths. (For example, on assembly lines, instead of one person doing the same repetitive task all day, people work in "pods" or teams that perform many different tasks as a group; the workers aren't faced with mind-numbing repetitiveness, and they can take more pride in their contributions to the finished product.) Of course, in a growing number of cases, these repetitive steps are now carried out by machines, which doesn't always lead to a happy ending.

How often have you tried to access needed information and heard those fateful words, "Sorry, the computer's down?" Or as happened to one of the authors recently: he found a huge "buy-everything-here" store ground to a halt when the checkout computer went down. No one could buy anything, and customers abandoned their piled-high shopping carts in the aisles to go visit the competition.[4]

Yet even when our machines and computers do work, they don't (at least yet) bring the "human" touch to what they're doing. Most machines always get four when adding two plus two, as do most humans, usually, but when circumstances require it, people have been known to sometimes get three-and-a-half, or perhaps five, solving problems in a way that our rather literal-minded silicon servants have yet to mimic.

But machines have continued to make our work "easier." For example, they're good at producing far more forms for people to fill out than mere people could produce. But historically to get the information off all those filled-out forms required huge armies of "key punch operators" to translate and keypunch the data into the room-sized information behemoths that predated today's computers. If some of that data had to be shared among different computers (which were generally independent computing islands unto themselves), then magnetic tapes would be hand-carried from one computer to another!

SKIPPING THE MIDDLE PEOPLE Today though, such human keypunching and tape-mounting intervention are needed far less often. Many computers are now networked together via private networks and the Internet to facilitate data sharing. Techniques for bringing information into our computers' insatiable maw have improved (for example, optical character recognition, where written or printed information is converted into text, and imaging of forms, where signatures, drawings, and other non-text material has to be stored for fast retrieval), and the World Wide Web has made it much more

---

[4] www.compaq.com/rcfoc/19990412.html#We_Are_At_Their

common for forms to be presented electronically in such a manner that their information is already digitized for error-free storage and rapid dissemination.[5]

The Eighties might be called the "reengineering decade,"[6] with an objective of achieving a quantum leap in overall company productivity. Feeding data directly into computers without human keypunching intervention, and networking, made the "care and feeding" tasks largely unnecessary, along with the people who used to perform those functions. Productivity, measured as the ratio between goods and services produced over the number of people active in this production, increased, because the volume of goods produced went up, although the number of people involved did indeed decline.

Consider a specific example. After investing millions in information systems, an American car manufacturer still needed 900 people to deal with invoicing and payments. So a project was designed to optimize related information systems, interconnecting them, simplifying how people interacted with them, and developing new financial applications. The impressive goal was to reduce the overall staff by about 300 people. Not bad.

However, before this new project was implemented, management decided to compare the result of their expected staff reduction with a similar company in Korea—and they found that the Korean company handled a similar invoicing and payment process with even fewer people. In fact, only ten people! How could they do this? Obviously, the difference had nothing to do with the productivity of the individuals, but it had a lot to do with the way those invoicing and payment processes were structured.

The American company received their invoices in one Payments office, where they were recorded. Then, that office contacted the purchasing department to verify that the contents of each invoice matched the order. The Payments office then contacted the receiving warehouse to verify that they actually had received the goods. Then, the production line was queried to verify that what was purchased actually worked correctly. The payment policies were then checked and, finally, the check was cut and sent out for payment. No wonder it took so many people!

---

5 Some countries are already allowing, and sometimes requiring, their citizens and businesses to file income tax returns electronically! Unlike in the United States, where citizens can only file electronically through intermediaries (and often for a fee), Brazil lets anyone file by turning in a floppy disk or connecting over the Internet. Similarly, New Zealand is now requiring that any business that has to pay more than $100,000 in taxes, file their forms over the Internet. This is profoundly changing the rules of the game (www.compaq.com/rcfoc/19990517.html#From_Out_Of_The).

6 Michael Hammer is considered the guru of reengineering. Some of his most important books include: "Beyond Reengineering," Harper Business, 1996, New York: "The Reengineering Revolution," Harper Business, 1995, New York; plus another interesting article, M. Hammer and J. Champy, "Reengineering the Corporation, Manifesto for Business Revolution," Harper Business, 1993, New York.

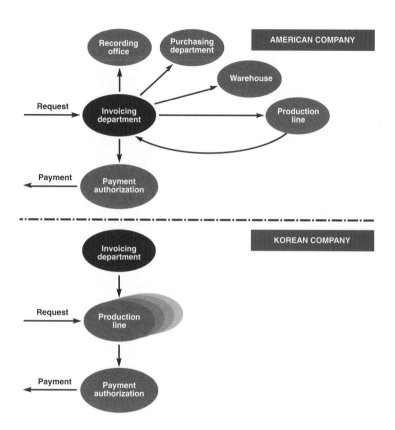

This comparison of the invoice and payment process between an American and Korean car manufacturer highlights the value of the significant decentralization of the second company. Such decentralization reduces process times and the required number of resources (600 people in the U.S. company vs. only 10 in the Korean firm)!

The Korean company's process was somewhat more streamlined, to say the least. The invoice was received directly by the production line, which had actually placed the order. They verified that receipts matched orders, and verified that those received items worked, in which case they released the check to pay the vendor. The ten-person central department only set overall payment policies and stepped in to handle exceptions.

The trick to reducing a 600-person department to 10 lay in stepping far enough back to view the desired results from an objective external viewpoint, ignoring "the way it has always been done." And that's a good lesson for us all.

Another example: Kodak used to have a product cycle[7] of about 18 months. At the end of the eighties, Kodak found itself being threatened by Fuji who had just changed the rules by introducing the disposable, one-time-use camera. Kodak needed an immediate countermeasure to avoid losing market share for its film and, perhaps as important, losing the public's perception of them as "the" name in film.

---

7 The time from when the decision is made to create a new product to when it becomes available on the market.

To address this need for a rapid response, Kodak created three teams, one each in the United States, England, and Singapore, but they weren't working separately on the project—instead they worked sequentially, as one team. At the end of the U.S. workday, the U.S. team would summarize the work they had accomplished that day and hand it off to the Singaporean team, who would shortly be coming in to work. When the Singaporeans got to work, they first reviewed what the Americans had done "yesterday," and proceeded to take the work farther. At the end of their day, they summarized again, and handed it off to the British team who again began by reviewing the work, doing their part, and summarizing for handoff to the U.S. group's workday. The end result was a project that took only four months compared with the expected 18 months, and most of the gains came not from the "follow the sun" 24-hour workday, but from the constant checking/reviewing/summarizing structure required by the constant "hand-offs" between separate teams!

Reengineering also extends beyond individual companies; it can include suppliers and even customers! Wal-Mart, now one of America's largest department store chains, radically reengineered how it put products on the shelves, getting rid of warehouses, their related costs, and much more. Wal-Mart simply provides display space for its suppliers, whereas it handles the products' image, sets and assures quality levels, manages the chain's image to its customers, and pays the suppliers for their goods that have been purchased.

But there is more—Wal-Mart does not tell its vendors when to restock their portion of the store's shelves—Wal-Mart's cash registers are constantly advising vendors how much of their products have sold, and the vendors are responsible for making sure that the shelves are kept stocked! Products go directly from the vendor's trucks to the store's shelves. Nothing is ordered. There's no invoicing. Business just flows according to preset supply and pricing agreements, oiled by the computing and telecommunications environment that knits them together into one virtual supply chain. Every possible constriction to this flow is simply ironed out of the process. Wal-Mart is free to focus on its most important process—the relationship with its customers. It isn't just Wal-Mart—similar processes have been adopted by both Fiat, at its Melfi, Italy production plant, and by the clothing manufacturer Benetton.

CLOSING THE GAP   Reengineering holds the promise of bringing customers and businesses closer together, by taking out traditional "middlemen." This is the often-discussed "disintermediation," where businesses that traditionally sat between a customer and the actual supplier are being rendered valueless as the Internet puts supplier and customer directly in touch. For example, those travel agents who primarily wrote tickets are on the wane now that people can purchase those same tickets, for the same price, directly on the Web.[8] On the other hand, how many people will go to each airline's Web site to search for the lowest fare? Few. This is why a new set of "intermediaries" has come into existence, usurping and replacing those Industrial Age ticket agencies that no longer add value. Companies such as Expedia and Travelocity[9] have taken over the cross-airline price comparison tasks of the old ticket agencies, and they go a bit farther; they're available 24 hours a day, and they offer personalization services that many traditional travel agents do not. The result is millions of dollars in ticket sales over the Web. These are still intermediaries, but they're Knowledge Age intermediaries, operating with the knowledge that their competition is but a click away.

Stock brokerages are another good example; the online discount brokerage houses began by offering rock bottom commissions on their transactions and also by supplying the Web's 24 hours per day ease of access. In just a couple of years, online brokerages have gone from zero to 1.2 million accounts (as of mid-1999), containing a half-trillion dollars in assets![10] These online brokerages are beginning to offer the "advice" that the traditional firms still felt justified their high prices; those traditional brokerage firms are now having to take careful notice indeed.[11]

There's another interesting fallout from this Knowledge Age trend of customers and suppliers getting closer together without the lubricant of middle men; many businesses that used to interact primarily with their distributors, who often spoke "their language," now find themselves having to learn to deal directly with their end customers. This is sometimes quite a shock, but it's definitely not a bad state of affairs.[12]

Another interesting change in how businesses now operate is the collapsing of the traditional "planning, implementation, and operation" phases of business cycles. For example, take cellular phone networks. Because of the overlapping nature of their geographic cells, if one cellular transmitter fails, the network automatically reconfigures itself to continue providing service, although perhaps with some diminished capacity. Indeed, even without failures, the

---

8 Lee Travel Group (www.leetravel.com) estimates that by the end of 1999, the number of travel agencies will decrease from last year's figure of 32,000, to 15,000, as result of the increase in ticket sales over the Internet. If it's true that the process of eliminating the middle man will cause some companies to disappear, it's just as true that new ones will continue to jump on the bandwagon. The point is that, in many cases, the value chain no longer considers simple distribution as value. However, the change paves the way to new opportunities to add value in the distribution chain. Therefore, on one hand, we will witness a progressive disappearance of middle men and, on the other, an improvement in brokerage skills with greater value for users.

9 www.expedia.com and www.travelocity.com

10 www.compaq.com/rcfoc/19990510.html#Ecommerce

11 According to Booz-Allen VP Grande Bucca in the April 27, 1999 Internet Daily, *"The only advantage left for the traditional firms is advice, and the online firms are closing fast in that area too."*

12 *"On the Internet, whoever owns the relationship with the customer, owns the future. . . If you want to get my attention, show me that you can help me save time."* by Gerry McGovern, May 9 New Thinking, NUA (www.nua.ie/).

Italian cellular phone company, Telecom Italia Mobile, redesigns its cellular network monthly based on changing traffic patterns. The traditional "Plan it, Implement it, Manage it" form of business has become one continuous process.

Of course "reengineering" is difficult, and not without its downsides, and in some circles it has developed a deserved bad name as being equivalent to downsizing, but the two can and should be very different things. Company cultures should change to make the best use of their talent pool and use the ripening fruits of the Knowledge Age[13] to find the best talent for the tasks at hand, wherever he or she may be physically located.

With all these Knowledge Age changes, the skills that businesses will value in their employees are changing as well.[14] People will still need their appropriate technical skills, but most will now also have to relate to customers. They'll have to be creative, innovative, and entrepreneurial, and they will have to be able to see the "big picture," rather than an Industrial Age narrow-focused "little cog in a huge machine" view. Because a businesses' customers will be much closer to each employee than ever before, each employee will be better able to make a difference and chart the company's course.

ALL FOR A GOOD CAUSE   Business changes demanded by the Knowledge Age won't be painless, but they will move, and potentially build wealth, at breakneck speed for those willing to play by the new "Internet" rules. As the United Kingdom firm Analysis said,[15]

"When the dust settles, lasting competitive advantage will lie with the companies that have managed to find a new way of doing business."

In this regard, how do you rate *your* business?

---

13 The explosion of the Internet is second only to the expansion rate of "corporate intranets." These are private networks based on the same communication protocol used by the public Internet, which are typically connected to the public Internet through a firewall that limits access and provides some security. Zona Research (www.zonaresearch.com) estimates that the global Internet market will be worth $43 billion in 1999, $28 billion will go toward supplying intranet equipment and services, whereas the remainder will be used for public Internet purchases. "Extranets," or secure extensions to private intranets to create a shared information environment to other companies (suppliers or customers) using the public Internet, are also on the rise www.Zdnet.com/intweek/daily/980310c.html

14 Some people may find their skills still valuable but to a different company, as an increasing number of companies "out-source" business functions that are not core to their business (such as payroll, warehousing, and delivery). This way the companies can more tightly focus on what gives them their competitive edge, while other companies focus on traditional tasks, such as payroll processing.

15 March 26, 1998, *PC World*—www.pcworld.com/cgi-bin/database/body.pl?ID=980326114855

2   Defeating Space and Time

A simple miracle, telephone calls, takes our voice to the farthest reaches of the planet in the blink of an eye. Today, high-quality videoconferencing can send our pictures as well, although it's still out of reach to the casual user. But that is destined to change within the next ten years as the cost of bandwidth plummets.[1]

What will the changes be once the transmission of images, video, and films becomes as easy and cheap as making a voice call today?

Let's take an example that's still making headlines: distance learning.[2] Typically using the Internet, students[3] can listen to (and watch) lectures, communicate with their teachers, ask questions, turn in their homework, and get feedback on their work.[4] Even students on a traditional campus can benefit from these facilities, which can also be used to bring guest lecturers into the classroom. An on-campus student could thus sleep in and catch the class at a more convenient time—the educational equivalent of delayed viewing in the world of television.

The ability to reach out and touch your education will have a far more profound effect, because the pace of change now requires that we all remain lifelong learners. Historically, school has been a normal phase in the choreography of our life. First we played, then we went to school, and then later to work, but due to the rate of change today, much of what we learned in school is becoming obsolete. Certainly in the sciences, where yesterday's state of the art is today's historical footnote, but also in areas like history, where discoveries shed new light on events that occurred ages ago; the textbooks we learned from no longer tell the full tale. We now constantly have to "stay in school" to refresh and update our general and technical skills. So what could be more timely and cost effective than to be able to take refresher courses from any convenient PC while we continue to carry on our personal and business lives?

1 In fact, it has already begun—in July 1999, Kyocera introduced a *wireless* videophone, the VisualPhone, which works on the Japanese PHS cellular network. Although not high quality and running at only two frames per second, these color images, in the palm of your hand, give us a hint of things to come (www.compaq.com/rcfoc/19990531.html#Another_Blow_To_Science).

2 For examples, see dir.yahoo.com/business_and_economy/education/college_and_university/business_schools__departments__and_programs/graduate_programs/distance_learning/

3 Distance learning is not just for traditional students but also for those who, due to location, illness, or physical limitations, might never attend traditional classes.

4 Of course missing an assignment might require new excuses. Let's see, the dog ate my floppy; or my Internet connection was down?

Speaking of textbooks and other "formal" references that no longer age gracefully, wouldn't it be better if our references kept themselves updated? Today, with "electronic paper" still in its infancy,[5] that isn't feasible, but a growing number of published works do provide updated information through links to a Web site. One good example is the CD-ROM encyclopedia, which typically now lets readers link with constantly updated information as they browse the CD.[6]

KNOWLEDGE IS NOT A SPECIFIC BIT OF KNOW-HOW BUT A NETWORK THAT EVOLVES OVER TIME

The definition of knowledge is also changing. It is no longer just what is stored in our brains, nor just what is written in books or amassed in libraries. Knowledge has become a fluid "state of the moment" in which the Web itself has become a fundamental part of that knowledge. Doctors will not be chosen purely on the basis of their experience, the number of books in their reference library, or how much equipment is available in the hospitals where they practice. Instead, that choice might be biased by what access they have to the network of global medical knowledge. Even our own illnesses[7] could become a part of this Web of knowledge to benefit others with similar problems.

Getting back to distance learning, let's look at ways in which it might help us become lifetime learners.

Let's assume that there are pools of experts around the globe who will give, or sell, instruction and knowledge in their areas of expertise and that there are people, also located all over the world, who are interested in those subject areas or need training. CD-ROM-based courses and the Internet are ideal media to bring these teachers and learners together.[8]

TEACHING AND LEARNING ARE DIFFERENT SIDES OF THE SAME COIN IN WHICH THE ROLES OF STUDENTS AND TEACHERS OFTEN OVERLAP

Examples of this kind of teaching and learning have already started to pop up. The GEMBA (Global Executive Master on Business Administration) program offered by Duke University (North Carolina) is enjoying tremendous success. Teaching is based on the use of CD-ROMs that allow students to interact, both locally and via the Internet, with their professors. The Internet is used to set up conferences between students and teachers as well as for discussion groups among par-

---

5 One company working in this direction is E-Ink— www.compaq.com/rcfoc/19990510. html#A_Sign_Of_The

6 This is another example where the Internet has changed industries. Encyclopedia Britannica had been the most popular and widely sold reference for the last 250 years. Yet the advent of encyclopedias on CD-ROMs knocked Britannica back to fourth place, with the first three spots held by CD-ROM encyclopedias. They gained popularity because they are far less expensive than a shelf full of books, and because through their Web links, they don't go out of date. Even the venerable Encyclopedia Britannica has had to bow to change, and is now also available on CD with links to its interactive Web site at www.eb.com PC World, in a May 1999 article (www.pcworld.com/cgi-bin/pcwtoday?ID=5154), also describes how now, in some cases, CD-ROM encyclopedias are in-turn giving way to free, full-content versions right on the Web!

7 With appropriate privacy concerns taken into account.

8 And perhaps it's high time for some of these changes. Consider if you suddenly had a guest from the eighteenth century. He'd be amazed by everything you did: switching on the lights, opening the water faucet, making a phone call, or watching television. Take him outside and his amazement would change to astonishment: an almost deserted countryside, but plenty of "out of season" fruit in stores, the cars, and the planes. There's only one place where he would feel as if he were back in his own century: the classroom. The same teaching styles, almost all the same topics, and many of the same tools would be there.

ticipants scattered around the globe and engaged in collaborative projects. So far, these technologies haven't significantly changed the teaching experience, only expanded it beyond the ivy-covered walls to a global classroom, but as bandwidth increases, and as innovative educational ideas are implemented, the potential is there to extend education in new directions.

BEING THERE

We waste a lot of time getting from here to there. An hour's commute to work. Time to drive between clients. Far too many "trips to the store." A day wasted to travel to a far city. If only we could put that time to better use.

That's one of the benefits that our evolving computing and communications environments promise to bring. No, the Star Trek transporters aren't just around the corner and may never come to pass, but scientists have already succeeded in teleporting tiny bits of information![9] On the other hand, even if the transporter never does materialize there are still ways to experience things at a distance.

BEING EVERYWHERE
WHILE STAYING PUT[10]

For example, suppose you wanted to explore the temples and plazas of ancient Egypt. There are countless books and online photos you can explore, but you're limited to just what the camera saw. True, you could go out and find a videotaped travelogue, but again you'd have to explore what its creator was interested in and at her pace. Today though, there's another alternative—"surround pictures." Essentially, a camera takes a series of 360-degree photos around, above, and below itself, and they are stitched together into a virtual picture that you can view over the Web.[11] Using your mouse, you can "turn around," look up or down, and zoom in or out to explore the environment at your leisure. Some areas allow you to "jump" from one vantage point to another, exploring one area from multiple perspectives or from a wider locale. Another variant allows you to stay in one place and spin an object around to view it from all angles.[12] One day, your computer might even automatically navigate you around such landscapes by using a TV camera to watch where you're watching and move your viewpoint accordingly!

But you don't have to wait for tomorrow—today, you can take a virtual taxi ride around New York, joining Clever Da Silva in his very real taxi cab on the streets of the Big Apple.[13] Or, you can shop for furniture for

9 In 1997, Anton Zeilinger and his team at the University of Austria (as well as a separate team in Rome) demonstrated that they could transmit the polarization of one photon to another, distant photon, and in 1998, CalTech physics professor Jeff Kimble used light beams made up of many photons to teleport information across about 3 ft. It's not Star Trek, but it is certainly a beginning!
www.compaq.com/rcfoc/981102.html#Beam_Me_Up_Scotty

10 For example, you can explore homes for sale by virtually "stepping inside" them over the Web—see www.ipix.com

11 The mysteries of ancient Egypt in Quick Time 3D format, via PBS—www.pbs.org/wgbh/nova/egypt/explore

12 Such as Colossus of Memnon—www.pbs.org/wgbh/nova/egypt/explore/memnon.html

13 See www.ny-taxi.com/

your den at Gallery Furniture.com from the comfort of your den, where you take control of cameras located around the store to tilt, pan, and zoom to look at all their wares.[14]

Here's an example of how the Web lets you be in two places at the same time: at the office, and keeping tabs on little Johnny or Suzy at the day care center. A growing number of such centers are offering Web-Cams that you can use to watch your kids during the day (they use security measures to assure that only the appropriate people can watch).[15] What? No kids? Don't feel left out—some kennels are putting in KennelCams so you can keep up on pet's vacation while you're at the beach.

These "virtual" technologies also hold the potential to create new markets. Consider, for instance, a work environment where many people "farm out" their services to a global pool of businesses on an as-needed basis. Virtual environments might provide a "job bazaar" more compelling than the many search-based "job classified" sites,[16] because you might be able to populate the virtual environment with samples of your work.

JOB OPPORTUNITIES AND JOBS ARE ON THE WEB

As temporary or ad hoc work becomes more prevalent, it becomes more difficult to establish a corporate legacy and lasting people connections. Online communities may take on an even greater importance in "networking" toward your next job.

VIRTUAL COMMUNITIES

Virtual communities began as text-based environments where people posted replies under given topics,[17] but they're now expanding far beyond these text-based roots. Some allow us to project our virtual self, in the form of an "avatar" or electronic body, into a world that represents familiar or surrealistic environments. In these, you can explore as if you were floating through an ancient or futuristic city—and you're not alone! Other avatars such as yours are there, controlled by other people somewhere out on the Internet, and you can each see each other's electronic bodies. The concept is sometimes called "telepresence," and you can chat with each other, cause your avatar to assume various expressions, and generally explore and carry on as if you were really there (but without the hotel bills and travel time, but, alas, also without the food). These are "virtual spaces," and

14 See www1.galleryfurniture.com/

15 www.parentwatch.com/pw/demonstration.asp/

16 Examples include employment.yahoo.com/, www.monster.com/, and thousands of others.

17 One of the first such environments was "Notes" from Digital Equipment Corporation.

although currently relatively crude, they represent the forerunners of virtual communities that will foster community, learning, and commerce.

Telepresence is about far more than visiting foreign lands (real or imagined). A variation called "telemedicine" is beginning to combine a host of technologies to bring medical treatment, not just medical advice, to places where doctors (or special expertise) don't happen to be. For example, the Department of Defense is exploring the idea of telemedical "battlefield robots" that controlled by surgeons behind the front lines (or at home in their hospital), can diagnose problems and even do surgery. Closer to home, surgeons in the same operating room as their patients are using similar surgical robots[18] to view the 3D environment within a body and to scale their hands' movements down to the microscopic distances necessary for working on tiny human structures. The robots can also filter out a surgeon's normal hand tremors that would otherwise prevent such delicate surgery.

Perhaps even more fascinating, such a robot can cancel out the movements of a beating heart so that the surgeon can make repairs without stopping it[19]; the surgeon sees the heart as if it were still, and the robot compensates as it moves the tools within the body! As the issues of controlling such robots over longer distances are solved, surgeons with specialized skills will be able to operate on patients that need their services without one or the other having to waste the time of flying there.

Of course not all telepresence robots will be used to put things back together—the Telepresence Rapid Arming Platform is a robotic rifle that a SWAT team or military force could use as the "front man" (front Robot?) in a hostage situation or firefight, protecting the human who is out of the line of fire while he controls it like a video game.[20]

Even while working in such a distributed environment, where the various players may never have met and live in distant locales around the globe, collaborative technologies can help the project move forward.

People can contribute ideas into the common information pool throughout the project regardless of formal meetings, when and where the ideas surface, and those ideas remain recorded for all to contemplate and enhance. People can work in parallel on the commonly available information, potentially speeding up the process.

---

18 See www.intuitivesurgical.com/r1system/body_r1system.html

19 See www.wired.com/news/news/email/explode-infobeat/technology/story/20920.html

20 TRAP (Telepresence Rapid Arming Platform) is a telepresence application in the military sector (www.infowin.org/ACTS/RUS/PROJECTS/ac214.htm) that uses a computer to aim a rifle. A local telescope and other detection devices are used to transmit images of the environment to a sheltered position where a soldier can aim a virtual rifle just like in a video game. The aiming computer makes all the necessary adjustments and controls the real rifle (see www.businessweek.com/).

BEING DEAD IS NO EXCUSE!

The adoption and expansion of these and related technologies open the opportunity for a vast new array of services that people will come to value.

For example, at Carnegie Mellon University[21] it is possible to ask questions of famous people like Lisa Sheridan and even those who are dead such as Einstein, and these "people" will understand the question and answer you in context! To make this possible, CMU researchers developed a program to interpret the question and query an extensive database of sentences and comments prerecorded by (or for) the famous person in question. The program uses those "contextual chunks" to construct a reply appropriate to the question, which is made to appear in character for the famous person being interviewed. As a result, you have the eerie impression that this person, perhaps known to be dead, is having a direct conversation with you.

So, for example, if you wanted to explore the theory of relativity, you might choose to do so with Einstein himself. What a great teacher to have.

Being dead is no longer an excuse for keeping silent! In fact, advances in computing plus brilliant execution by some of today's inventors now have the dead continuing to produce new work!

At the end of 1997, music experts listened to a composition by Bach, described to them as an unknown passage that was only recently discovered. The experts immediately recognized Bach's style and melodic patterns. However, that piece was artificially created by a computer program that had been taught the "style of J. S. Bach." The piece was a forgery with respect to Bach, but an original with respect to his style, the result of work by Professor David Cope at the University of California who developed a system to analyze musical pieces and create new ones based on the same musical style.[22]

(Which begs the question, who owns these new, old works? Although that may seem academic in regards to someone dead for centuries, what if this program were to analyze the work of current artists and produce new songs that were recognizably their style? Are our legal and commerce systems prepared for such things?) Who knows, further advances might let you interactively jam with Louis Armstrong or the Beatles, or compose verse with Emily Dickinson, or—well, the possibilities are endless. And eventually, when you're in an electronic "chat" with someone, you may find

---

21 See www.grandillusionstudios.com/webdemos.htm

22 www.cio.com/archive/011598_trendlines.html

yourself wondering if the person on the other end is a he, a she, or a silicon it.

The Knowledge Age is already beginning to provide easy access to the intelligence and knowledge of all of its participants—people and not-so people, alike. As time passes and technology makes more miracles possible, what would it be like to live in a world of quite intelligent "things?" Let's take a look.

Communication Infrastructures

When we use a telephone for a "classic" voice call or we connect to the Internet with our PCs, we are making use of an enormously complex set of resources called the telecommunications infrastructure, which transparently makes a phone call appear effortless. In this section we'll get a better understanding of what's "under the covers" of that seemingly simple telephone (or modem) connection.

Lets begin with an analogy. The roads we use to drive from our homes to our offices (assuming we're not telecommuters) are part of a shared infrastructure, a common resource we all use for transportation. But the roads don't necessarily represent the most direct way for any one of us to get to work; the distance we travel is usually farther than if we used a helicopter to "fly" directly there. That's because the design of the road system had to work around natural barriers such as rivers and mountains, and artificial barriers such as existing towns and villages. The design of each road also had to consider expected traffic loads; high-capacity highway systems feed into smaller roads, which themselves are sized according to their expected traffic. The result is an infrastructure that can be widely shared by as many users as possible but isn't optimized for any one user (most of us don't have a multilane Interstate highway directly between our home and office). Nevertheless, this generalized "transportation infrastructure" has (literally) changed the face of our world, and of how we each work, live, and play.

This is wonderfully similar to the evolution of the telecommunications infrastructure in the twentieth century. In fact, there is a new phenomenon that we'll be exploring throughout this book: the synergy, or mutual influence, between physical (transportation) and virtual (telecommunication) infrastructures, which can be seen at two levels:

- The transport of quantities of people, goods, and services over our physical infrastructures is more and more enhanced by technologies in the virtual infrastructures. At the same time,

- as a growing number of products and services (such as music) can be purchased and delivered in both "realms," we are seeing fundamental shifts in how and where people buy.

This chapter takes a closer, more technical, look at the evolution of the telecommunications infrastructure, and at the "backbone" networks that are widely shared by many telecommunications services and users—very much like our physical highways are shared by many transportation services and cars and trucks.

MY VOICE IS BREAKING UP! The telephone network has evolved over the last 100 years to reach out and touch the most distant points on Earth, and it did so, initially, using copper wires and mechanical "central office switches."

Given the significant cost for each mile of wire and for each switch, it was a natural evolution to charge for calls based on the distance (the number of miles of wire and the number of switching offices traversed) between the caller and the person receiving the call.

Similar to how the cars, trucks, and buses of many businesses can share a single highway, technology has allowed us to carry an ever-increasing number of conversations over the same copper wires. For example, PCM (Pulse Code Modulation)[1] allowed us to bundle, or "multiplex," many conversations over a wire that used to carry only one call. Essentially, speech is sliced and diced into tiny 125-microsecond segments, or 8,000 samples, each second. Then, the volume, or amplitude, of each of those 8,000 samples is measured, and the resulting value is stored as an 8-bit digital number—one of 256 possible "volumes." It's that number—actually 8,000 of them for each second of conversation—that's sent down the line. Strange as it may seem, these numbers (reasonably) accurately represent our voice.[2]

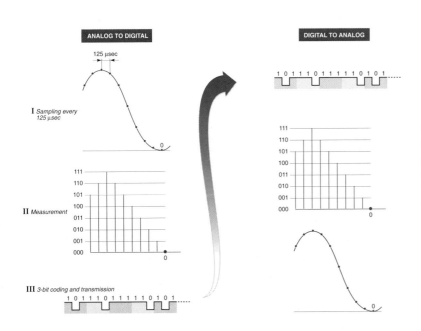

These graphs illustrate the mechanism used to digitize and convert an analog signal. For the sake of simplicity, we used a 3-bit code (8 levels), whereas the PCM system uses an 8-bit code (256 levels).

---

1 A technology based on studies by Reeves back in 1938.

2 This "analog to digital conversion" marks the true beginning of the digital revolution in telecommunications. This conversion originally required expensive equipment, and so for many years it was applied only over routes carrying the most traffic, so that the equipment costs could be amortized over many calls. By the 1980s, however, as the equipment costs continued to decline, most of the U.S. telephone network backbone (central office to central office) was digitized. Later, with the advent of ISDN (Integrated Services Digital Network) and later DSL, "digitization" could even be carried beyond the telephone central offices, right to the users.

So, having encoded the speech into 8-bit numbers and needing to send 8,000 of these numbers each second, it turns out that we need a transmission speed of 64,000 bits per second, or 64 kilobits per second, to carry each direction of a phone call. But as phone traffic grew, we could no longer afford to dedicate a physical wire for each phone call[3]—that's where "multiplexing" came in.

In the 1960s, we were able to transmit 1.544 million bits per second over a single pair of wires (called "T1"), so this made it possible to transmit 24 calls[4] per pair. In a move that was only the very beginning of what has proven to be a continual process, overnight the transmission capacity of every pair of wires (potentially) increased by 24 times!

THE PHOTON TRANSPORTATION SYSTEM

Even with subsequent advances in multiplexing technology, we were reaching practical limits as to how much data (once digitized through PCM, our voice calls were really data) we could pack into a pair of copper wires, and even as the end neared for increasing the bulk transmission over copper, innovative scientists found that "light pipes," or hair-thin glass threads, could carry more data than copper had ever dreamed of.

Beginning in the 1980s, fiber began to replace copper (a good thing, since many cities were again running out of underground room for ever-more thick copper cables), and, like copper before it, we just keep getting better at packing more data into each and every fiber.

The basic data rate we can push down today's fiber is 10 gigabits per second, but a "mere" 10 gigabits per second doesn't nearly make use of a fiber's multiterabit (a terabit is a thousand gigabits) theoretical limit.

One technique increasingly used today to expand a fiber's capacity is called "Wavelength Division Multiplexing" (WDM), which works like this. If we can send a 10 gigabits per second data stream down a fiber using red light,[5] it turns out that we can also send a completely separate 10 gigabits per second stream of *green* light down the same fiber at the same time! Similarly, we can also send streams of yellow, blue, and other colors of data—all at the same time and without interference! At the receiving end, the individual colors are sorted out and their data separately turned back into electrical signals. It's almost like getting something (vastly more bandwidth per fiber) for nothing! Indeed, in early 1998, MCI began using an eight-channel WDM fiber to carry 80 gigabits per second.

---

3 The decision to use PCM and its multiplexing technology was made in New York following a ban imposed by its Mayor—telecommunication companies were forbidden to dig new trenches to add cables, although demand for new phones and call traffic were increasing! In retrospect, PCM might not have been the best technology to use, but it was the only effective method for increasing capacity right then.

4 In Europe a slightly different multiplexing, or bundling, standard was chosen: 2.048 million bits per second links containing 30 channels for voice plus two for signaling.

5 The "colors" actually used aren't in the spectrum that our eyes can see, but WDM is easier to understand if we discuss it in terms of colors, or wavelengths of light that we're familiar with.

Today, according to Lucent CEO Rich McGinn,[6] WDM already makes it possible to carry up to 80 full-speed data streams on a single fiber. By 2000, MCI expects to be able to pack 128 "colors" of light into each fiber, yielding 1,280 gigabits per second (or 1.28 terabits per second) of data. Competition being what it is, in May 1999, Nortel Networks announced 160 channels of 10 gigabits per second data over a single fiber, or 1600 gigabits per second (1.6 terabits per second), to hit the market in 2000![7] Of course these are slow compared to what's cooking in NTT's Tokyo labs—a fiber system that boosts each "color's" data rate from 10 gigabits per second to 160 gigabits per second and packs 19 of these streams into a single fiber, yielding an incredible 3 terabits per second of data through one hair-thin glass strand[8] (and we can be sure that even this is just the beginning)!

Lucent's McGinn further predicts that Dense WDM (DWDM) techniques will raise this to *"thousands of wavelengths in only a year or two."* That means that a single fiber could carry *"more than a terabyte per second."* Putting that in perspective, from Nicholas Negroponte's book "Being Digital," just one of these hair-thin DWDM fibers could transmit every issue of the *Wall Street Journal* ever printed—in less than one second, and because forward-thinking companies have been burying LOTS of fiber strands, this means we could have a communications backbone with so much capacity that even the exponentially growing Internet won't run out of bandwidth—for awhile. (Consider how far we've come—from Samuel Morse's first telegraphic phrase "What hath God wrought" sent over one strand between Washington and Baltimore at 5 bits per second, to, soon, a thousand billion bits per second—in 150 years. It boggles the mind.)

But all of this raw capacity isn't enough. The other part of this equation, the ability to route this incredible amount of data to and from the right places, still needs work. The way this is commonly done today, at network junction points where the fibers come together, is to translate the light back into electrons and run them through conventional (but ultra-fast!) routers, then converting the signals back into light for their trip to the next junction point. That's very inefficient and slow, and the speed of data on the fibers is beginning to approach the theoretical 50 gigabits per second limit for electronics. So are we stuck? Of course not. Consider if a data packet could be converted into light early in its life, *stay that way* as it's optically routed across the hugely dense fiber mesh, and only be converted back into electrons at the end of its journey. The capacity of our networks would be huge, and the latency (the time it takes a packet to get from here to there) would shrink,[9] but that's exactly where "All Optical Networking" is headed.

---

6 April 15, 1999, "Computergram International."

7 See www.nortelnetworks.com/corporate/news/newsreleases/1999b/5_4_9999280_OPTera1600G.html

8 See www.techweb.com/se/directlink.cgi?EET19990329S0024

9 A fascinating, easy to read explanation of how this works is at www.lucent.com/context/opt_net/hot5.html (hint: it's all done with nanotechnology mirrors), and you can find additional insights into these and related areas of optical networking at www.lucent.com/opticalnet/tech.html

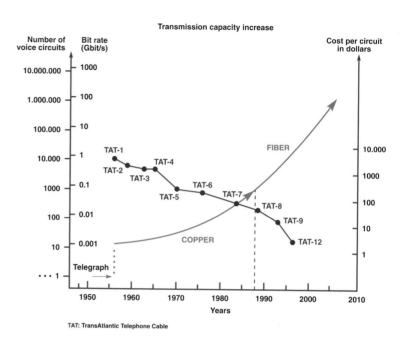

The diagram illustrates the enormous (100-fold) increase in transmission capacity between 1955 and 1985, which will increase another 1,000 times between 1985 and 2005, thanks to the use of optical fibers. The overlapping graph of TAT (transatlantic telephone cables) shows the decrease in cost of a single circuit, almost 10,000 times less than in 1955.

A SWITCH ALONG THE ROAD

Communication infrastructures have their roads (the wires and fibers), but as we were just discussing, they also have intersections through which traffic routinely must be rerouted. In telecommunications, these intersections are known as *switches,* and they're used to decrease the number of wire, and now fiber, "roads." Similar to the airlines' "hub and spoke" concept, which has us making connections to get from here to there, instead of requiring a wire to run between each and every point on the network, the telecommunications infrastructure is set up so that, just like an airline trip from Portland, Maine to Portland, Oregon, a data packet might take its first wire to a switch in Boston, which puts it on another wire to Chicago, where another switch moves it to a final wire to Portland, Oregon. (We do, though, wonder if the packet gets served better food on the fibers than we do on the airplanes.)

Because of the large number of switches around the country and because special "repeaters" had to be installed every few miles along each cable between switches, the farther a call went the more of this expensive infrastructure it used; that's why long distance calls used to cost more the farther away we called. Today though, thanks in part to the very long distances fiber can carry a signal without expensive repeaters and also due to the tremendous bandwidth that these fibers offer (so there's plenty of "extra room"), it doesn't much matter how far your call goes—the cost per minute, at least within the United States, remains the same.[10] (In fact, your

---

10 Indeed, this is spreading—as of July 1999, all calls within Norway are considered "local."

call from Boston to Miami may well find itself routed through Los Angeles if that's the best route open at the moment; it doesn't cost the phone company, or you, anything extra.) Even internationally, where we still sometimes pay far more than the typical United States 5 cents per minute, it really isn't distance that determines the cost; it's much more a matter of treaties or contracts or politics. For example, these days you can call from the United States to the United Kingdom for about the same price as a call within the United States, although a call to the geographically closer Iceland is far more expensive.

The Internet, as well, will happily route the data packets that make up your communications along many different paths to their eventual destination; in fact, the many packets that make up a single Email message, or an Internet phone call, may take wildly different paths, but be seamlessly reassembled into a single message, or short bit of sound, at its destination. (Remember that the Internet had its roots in a military communications network; it was explicitly designed so that the failure of any single route or "switch" would not keep the packets from getting through.) Although that "path redundancy" provides a very robust network, it also makes it impossible to "guarantee" a consistent route for all the parts of a message or Internet phone call. Although this is no big deal for an Email or fax message, for real-time communications such as voice and video, it introduces quality issues that have left "free" Internet phone calls with various quality problems. (By comparison, the traditional "circuit-switched" telephone network reserves a constant circuit between the caller and receiver for the duration of the call; that's how it's able to deliver higher quality.)

MAKING WAVES So far, we've been discussing the "fixed network" of wires and fibers, but as everyone who has used a cellular phone knows, radio is another way to reach out and touch someone. Indeed, why don't we just do away with the very expensive method of burying wire and fiber, and do everything wirelessly?

The problem is limited radio spectrum, which, using current technology, limits the total amount of information we can send through the air.[11] My first mobile telephone, years ago, took up half my car's trunk and had a maximum of four channels—that meant there could be only four simultaneous mobile telephone calls in the entire city! Over the years, as we found ways to use higher frequencies (which have different propagation characteristics) and cellular techniques (which reuse frequencies in small cells around a city), we've been able to make cellular phones affordable to most anyone. Indeed, we've also moved such "cells" into the sky. Motorola's Iridium satellite-based phone system, which became operational in 1998, inverted the idea of cells. Instead of the cells sitting fixed in one location (the cellular phone company's

---

11 There are, of course, some technologies that hold at least the promise of changing these rules to provide all the wireless spectrum we could use—Ultra WideBand (UWB) radio is one (see www.compaq.com/rcfoc/19990419.html# Science_Fiction_Falls).

tower), the cells have been placed in many satellites orbiting the Earth. In a land-based cellular system, as a user moves away from one cell and enters a closer stronger cell, the system "hands off" the user to the stronger cell, invisibly instructing the phone to change to the stronger cell's frequency. Similarly, in a space-based system, while the cells themselves (and the "footprint" they cover beneath them on the Earth) are moving far faster than the user (we hope), the technique is the same—as the edge of one satellite's cell approaches the user's location, the user is switched to the next, slightly overlapping satellite's cell. Iridium is but one of many such systems that are planning to take to the sky[12]; next, Globalstar is scheduled to become operational before the end of 1999. (Unlike Iridium, which routes the calls between satellites and then to the earth-bound phone system as close as possible to the destination, Globalstar picks up the call and immediately sends it back to the terrestrial phone network as quickly as possible.[13] There's also a sky-based system optimized for (slow) data traffic— Orbcomm,[14] operational since July 1998, which provides real-time Email services, but mainly for "things" rather than for people. For example, a trucking operation may use Orbcomm to keep in touch with its far-flung trucks, picking up and delivering information, new stops, etc., to each of its trucks.

But when it comes to high-speed data, we're going to have to wait until about 2003 when Teledesic gets off the ground.[15] This brainchild of Bill Gates and Craig McCaw will use 288 low earth orbit satellites (LEOS) to provide very high speed (64 megabits per second) Internet In The Sky to any point on the surface of the Earth.

THE INSATIABLE DEMAND
FOR BANDWIDTH

Lots of choices! But do we have too *many* choices for how to send our voice and data from here to there? Will we use all the communication media at our disposal? Or is technology providing more bandwidth than we can reasonably be expected to consume?

Some estimates, based on the growth of data traffic generated by the Internet, predict that traffic will reach the level of petabits per second (millions of billions of bits per second) by the year 2002.[16] These numbers are difficult to comprehend, but to get a sense, one petabit per second would correspond to 660 million television channels or would allow each person on the earth to download the full Encyclopedia Britannica from the Web every day. Clearly ridiculous, right? We'd *never* need that kind of bandwidth, would we?

---

12 Interestingly, Iridium (www.iridium.com) turns out to be the first "entity" assigned its own country code (008816), just like a country, but a country whose territory covers the entire globe.

13 See www.globalstar.com

14 See www.orbcomm.net/

15 See www.teledesic.com

16 We already have the names for speeds beyond the petabit: *exabit* (a billion billion), *zettabit* (a thousand billion billion), and *yottabit* (a million billion billion). Assuming a steady traffic growth similar to that created by the Internet over the last 3 years, we could actually be consuming a yottabit of bandwidth by the year 2015!

BUT IT WON'T JUST BE PEOPLE   Let's look at this from a slightly different perspective. Imagine that the users of this vast new Internet, the consumers of this enormous amount of communication bandwidth, will include not only people but also a myriad of "things," from vehicles receiving up-to-the-minute digital updates about traffic and road conditions, to schools exchanging educational programs, to industries synchronizing their production based on ongoing real-time information provided by their suppliers and customers, to cash registers in large department stores communicating directly with the companies stocking their shelves to ensure that they'll be replenished as required, to vending machines calling when they need to be refilled. This opens up so many entirely new, potentially valuable ways to make use of communications that, all things considered, such "absurd" amounts of bandwidth do indeed begin to make sense.[17]

Remember that massive amounts of fiber alone won't make everything work in the "hub and spoke" concept of the Internet; we also have to be able to "switch" this vast amount of data to get each packet where it wants to go.

Research is underway to deliver switching systems, the size of a beer can, that can switch a thousand billion bits (a terabit) per second.[18] (By comparison, it's amazing to realize that in the 1960s a central office exchange, which could switch only a few thousand million bits per second (a gigabit/second), required a million times more space! Or that your pocket cell phone contains a computer that, just 20 years ago, would have occupied an entire room!)

Looking further into the future, the distinction between transmission fibers and switching may gradually disappear. These two building blocks of modern networks evolved out of the need to optimize the use of scarce network resources. The more switching nodes that were inserted in the network, the better use we were able to make of the wires' limited transmission capacity. Conversely, the more transmission capacity we have, the fewer switching nodes we need. Taking this to the extreme, if we could have unlimited transmission capacity (or at least far more than the capacity required to satisfy our demand) then we could have a purely transmission-based network (no switches at all) with all users connected to the same wire.[19]

---

17 Consider this, in 1996 Internet traffic on MCI's network took up only 1% of its fiber network capacity; voice traffic covered the other 99%. In 1999, data accounted for 40% of MCI's traffic, and in 2000, MCI estimates that data traffic will exceed that of voice traffic!!

18 Stanford University (United States), Tinytera project—see tiny-tera.stanford.edu/tiny-tera/index.html

19 This isn't entirely fanciful—the concept has been considered in terms of getting rid of databases. If we had enormous transmission capability, then all the data we wanted to "store" could be placed on a transmission ring where it would circulate continuously. When someone needs some of the data, all they would have to do is connect to the transmission ring and wait for the data to pass by. With enormous bandwidth the waiting time would be very brief, and thus a traditional database wouldn't be needed at all! Indeed, this method of sending information around and around as a form of storage is being used today in "All Optical Switching" to keep a data packet waiting if necessary—a 30-centimeter length of fiber introduces a delay (and thus a storage time) of 1 billionth of a second.

SOFTWARE MAKES IT ALL WORK

This evolution of communications networks is made possible by advances in electronics, optoelectronics, and radio engineering, and by the digitizing of voice into a form that computers, including new telephone company switches, end-user PCs, and specialized "appliances," can manipulate. Because it's now really computers that do the telecommunications work, software has come to play a critical role. Those millions of lines of computer codes are the "glue" that allows us to layer applications and services on top of the basic network infrastructure and ensure their correct operation.

We, as users and consumers of telecommunication services, do not normally see, or need to see, the machinery and the plumbing of these networks; in fact, if something doesn't work, we'd generally prefer to place one call and have the problem, whatever part of the infrastructure is causing it, identified and resolved. If it's done right, the end user neither knows nor cares which supplier, or even which network, might be carrying his or her voice or data. Indeed, when it works, the global communications infrastructure virtually disappears from view; it falls beneath our notice, and that's a good thing, because already today, and far more so tomorrow, if the global telecommunications infrastructure were to fail for any period of time, a global business and social crisis would surely follow.

THE PATH AHEAD

Of course there is far more to the evolution of the transmission and switching elements that make up our modern telecommunications networks, but we've tried to give you an overview of where we are, how we got here, and where things might be going, as well as a view into a future that is likely to contain one, seamless, global, very high bandwidth, digital network that makes no distinction as to what type of information it's carrying. What the end-user devices send will be packets of bits, which, depending on the device or its whim, might be an Email message, a phone call, video from a videoconference, music, or a collaborative document that many people can operate on at once. Or—entirely likely—with such a flexible network, the bits may represent some entirely new class of information we haven't yet dreamed of! Gee, this sounds like the Internet, doesn't it.

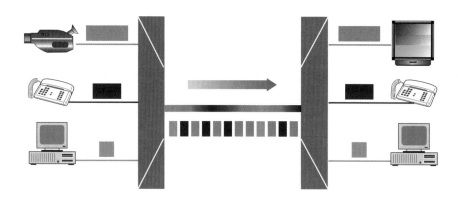

The diagram depicts the ATM coding and transmission model. The information coming from cameras (video), telephone (voice), and computers (data) is broken down into packets with the same length (53 octets: 48 octets + 5 octets for the address) and sent over the same line. Upon arrival, each data stream is separated and sent to its own destination.

3   Living in a World of "Intelligent Objects"

I'M INTELLIGENT, BECAUSE I MAKE THE RULES!

Both carbon and silicon atoms can bind with other elements in a myriad of stable structures, but nature bet almost everything on carbon—we and almost all living things are carbon based (although we recently have found just a few sulfur-based single-cell organisms living far beneath the ocean, clustered around hot volcanic springs). Semiconductor scientists, on the other hand, prefer silicon because of its properties as a semiconductor.[1]

Fundamental to our conception of "intelligence" are the abilities to receive, process, and transmit information, which can be done using both carbon- (we, after all, are living proof) and silicon-based structures. However, "intelligence" must also be able to lead us beyond simply carrying out programmed processes (such as the movement of a clock or a computer program).

The 5,000 cells in a snail's brain allow the snail to distinguish between a tasty leaf of lettuce and another leaf that's poisonous. The brain cells of a snail are, from a functional standpoint, identical to those in our brains, so why can't the snail compose a sonnet? Similarly, scientists at the University of San Diego have developed a working computer reproduction of a snail's brain, an emulation based on neural networks,[2] which can also distinguish between good and bad leaves, without the scientists having to teach it about lettuce or what makes it "good." That's a pretty impressive achievement, but this neural network still hasn't written a sonnet (that we're aware of). So, are "intelligent" computers right around the corner?

Stating that Deep Blue doesn't think because it utilizes schemes that differ from our brain, is like saying an airplane doesn't fly because it doesn't flap its wings.[3]

1 Silicon is a pure semiconductor. At room temperature, its resistance is between that of metals and insulators and can be controlled by the selective addition of impurities. It's the foundation material of the computer industry and of most of the electrical and optical infrastructures of the Knowledge Age that we describe in this book.

2 Neural networks are software that attempt to reproduce brain structures and simulate thought processes, specifically with respect to learning and pattern acquisition; this should also be looked at as an emulation of learning, because we do not really know which brain structures are involved in learning and how this happens. Neural networks could just be software machines that involve different structures than what humans use for learning, with similar results.

3 Statement made by Drew McDermot, Computer Science professor at Yale, talking about the IBM computer called Deep Blue that defeated Kasparov, the world chess champion, in 1997.

Our brain contains a few more cells than the snail's (about 25 billion more, give or take), with an immense range of interconnections. It's that huge number of cells, plus their vast number of interconnections, which scientists think provides the "something extra" that leads to our (sometimes) intelligent behavior.

On the other hand, our arm has a similar number of cells as our brain, but most of the cells in our arm lack the functional specialization of our brain cells, or the structures needed to tie them together into a dense matrix. So, simply the number of cells does not ultimately light off the spark of intelligence.

The ability to reason seems to require an interconnection framework, the more sophisticated and structured[4] it is, the more sophisticated and variegated the resulting thoughts. Once a certain threshold is reached (and we have a problem defining that threshold), we achieve thought and intelligence.[5]

In addition to the huge number of cells, our brains have an interconnection infrastructure of trillions of connections, and this combination does (usually) result in intelligence. We've already created computers with tens of billions of "cells,"[6] but they don't think. Why not? One difference is that this computer only had around a billion interconnections, far fewer than in a human brain.

Because conventional computers have far fewer "cells" and interconnections, it's no wonder they don't (yet) think. However, instead of using billions of "wires"[7] to send small, slow bits of information between cells in a brain, computers use just a few wires that carry a tremendous amount of information per wire.[8]

Microelectronics keep evolving at an incredible pace, which means that more of the objects around us are gaining the ability to store, manipulate, and communicate information. Even collections of these devices, or networks, have not yet become intelligent, perhaps because there aren't yet enough interconnections to pass that elusive "intelligence" threshold, but as more and more devices interconnect at faster rates,[9] we have to wonder at what point might we have "enough" silicon cells and "enough" interconnections between them.

4 "Climbing Mount Impossible," Richard Dawkins, W.W. Norton, 1996.

5 The problem in defining intelligence is qualitative by nature, because we don't have a formal definition of intelligence or of the ability to think. Usually, we tend to define intelligence in our own terms, if something can be done only by a human being then we define that as a sign of intelligence. The problem, of course, is that even today's "simple" computers are forcing us to slide the threshold point. For example, the ability to play chess "well" used to be considered a sign of intelligence—until IBM's Deep Blue beat human chess champion Gary Kasparov. Suddenly, we decided that intelligence really wasn't necessary to play chess—only the ability to calculate very fast and run many future "decision trees" is needed. What "intelligent" activity will fall to our silicon friends next?

6 See mission.base.com/tamiko/cm/cm-text.htm, a reference to one of the early massive parallel computers, the MIT Connection Machine.

7 The wires we have in our brain are called *dendrites,* and there are about 1,000 for each brain cell.

8 Taking a simplified approach we can say that interconnection power is based on the speed of messages, around tens of millions of messages per second in a computer, but only 5 to 100 messages per second in nerve cells—multiplied by the number of available lines—hence the interconnection ratio of 1 to 10,000 between a computer and our brain.

9 Research firm ARC Group expects that by 2004 there will be a billion people connected to the Internet, and three-quarters of them will be doing this over handheld wireless devices running at 2 megabits per second—www.compaq.com/rcfoc/19990614.html#Convergence_And_Kids

ACADEMIC THINKING   Now that we're thinking about "thinking," let's think about MIT's research project called "Things That Think."[10]

Imagine that we're living in an environment where most of the objects around us have computing power (call them "Information Appliances" or, generically, "communicating, computing appliances"). They have knowledge of the local world around them (through sensors) and of the broader world beyond (through the Internet).[11]

Household objects, from the doorbell to the oven, could know whether or not we are home and react accordingly. For example, if someone rings the doorbell, the picture from its camera could automatically be presented on the display screen in whatever room we're in.

Since 1996, more PCs than television sets were sold in the United States. Experts predict that by 2005, more information appliances than PCs will be sold, and that by 2010, they will outnumber personal computers by a ratio of 10 to 1.

Using voice commands, so we don't have to get up and push a button or hunt for The Remote (*"Computer, let me talk to the person at the door."*), we will be able to start talking to whomever knocked and even remotely open the door if we so choose. If we're not at home, this voice and video and remote control transaction could still take place through our cell phone, allowing us, for example, to remotely open the garage door for the delivery of a package![12]

DIGITAL FASHION   Let's examine some intelligent objects that were explored in a cross between a fashion show and a technology exhibition put on by the likes of MIT's Media Lab, Nike, Swatch, and Levi Strauss. There was a cell phone that had no buttons—all dialing was done by voice; ultra-light microcomputers that were woven into clothing, into the wristband of a watch and into the frames of a pair of glasses; a 3D glasses-mounted virtual display, using two lasers mounted on the edges of the glass lenses to project images on the retina[13]; sensors embedded in garments that automatically check up on your health—and, at least these days, a belt full of batteries to power all these goodies.[14]

10 The Things That Think is an MIT research project: www.media.mit.edu/ttt/

11 For example, "smart materials" are being developed that react to the world around them, such as "smart concrete." (See www.abcnews.go. com/sections/science/DailyNews/ carbon_electricity0304.html). It contains carbon and insulating structures that allow it to sense its environment (such as strain) and to pass on "interesting" information. In some cases, smart materials can also positively react to what they sense, such as "smart skis" that actually damp vibration. (See www.forbes.com/asap/98/0601/ 048.htm).

12 Do you think that video cell phones are science fiction? Kyocera has already released a video cell phone in Japan, www.kyocera.co.jp/ frame/product/telecom/english/ vp210_e/index.htm They expect to sell 50,000 the first year. Of course the bandwidth available to today's cellular phone is too narrow to carry a television signal (Kyocera only transmits two frames per second vs. TV's typical 30). However, higher wireless bandwidth will soon be available! For example, "General Packet Radio Service" (GPRS) is a technique that will initially provide 115 kilobits per second of wireless, always-on, Internet connectivity, and other "3G," or third-generation cellular phone techniques, promise 2 megabits per second of wireless connectivity within a few years. Even tiny screens, such as on Kyocera's video cell phone, will get bigger—virtually. Specialized optics sitting on the end of a phone's "flip" can project what appears to be a 21-inch virtual screen right into your eye! See www.siliscape.com/ our.htm/

13 See www.compaq.com/rcfoc/ 19990503.html#Default_13 and see www.microopticalcorp.com/

14 But that may not be necessary in the future. MIT is working on a button-sized turbine generator that may address our portable power needs. See raphael.mit.edu/ breuer/pubs/ MEMS99_MicroBearing.pdf

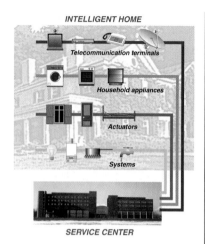

INTELLIGENT HOME

Telecommunication terminals

Household appliances

Actuators

Systems

SERVICE CENTER

15 See wearables.www.media.mit.
edu/projects/wearables/out-in-
the-world/beauty/index.html for
ideas on how fashions and
technology may be coming
together.

16 See www.wired.com/news/news/
email/explode-infobeat/
technology/story/20748.html

17 See the "dancing shoe" at www.
media.mit.edu/physics/
danceshoe.html

18 Sony registered the "Netman" trade-
mark in 1997 for a system that
should be available in 2000, which
will keep us constantly in touch with
the global information network. The
revolution of the Walkman, which
put music on our belt, will probably
be surpassed by Netman-like
devices, which will put each of us in
touch with the Internet no matter
where we are.

19 In February 1998, a medical team
headed by Prof. John Perkins
(director of the Clinical Engineering
Department at London's Imperial
College) gave Tony Blair, U.K.
prime minister, a sample microcom-
puter that could be inserted under
the skin to monitor various health
criteria. If a potential problem were
detected, it could trigger an information feed to his physician, perhaps even reserving a hospital bed!

Such fashions[15] are not just destined for the office.[16] We already have experimental clothing that detects movement and converts it into music. Instead of dancing to the rhythm of a song, a performer wearing these clothes actually creates the music by dancing![17]

These devices, which are "worn" as clothing, can communicate with each other, using the wearer's skin as a conductor (which is actually quite efficient and safe at high frequencies). The possibilities offered by such articles of clothing are limitless! And the cell phone, of course, can be their gateway to the outside world.[18]

True, this equipment may seem a bit cumbersome right now, but over the next five to ten years these ideas may become "old hat." For example, when was the last time you noticed the buttons on your shirt, which were themselves once a significant innovation?

JUST NEAT IDEAS? But are these just "neat" ideas for the techno-student? Perhaps at first, but as they develop and mature, they hold the potential, like the Walkman, to change our lives. For example, consider health services. Chronically ill patients could be kept under constant surveillance by medical centers' computers.[19] If problems arise, it may be possible to take immediate corrective action by remotely causing an implanted "chip" to release specific drugs.[20] Patients with heart disease might be able to live a completely normal life, knowing that any problem would be detected and reported as soon as it occurred and then easily controlled from the remote medical center. Support services for the disabled would become more effective with the availability of bioelectron-ic transducers[21] that can treat a wide variety of hearing, visual, and per-ceptual defects. Even common clothing, like the bra, may soon get a tech-nological assist.[22]

Other types of help we can expect from all these "smart" devices include never getting lost,[23] and help in avoiding losing those around us (such as our kids).[24] But having smart things know where we are can also help us to find the nearest gas station, or newsstand, or Italian restaurant. The smart things we're already carrying will provide these and other new services as a matter of course.

ON THE FARM   Speaking of knowing where things are, GPS (Global Positioning System) satellites are already helping out "down on the farm."

"Focused agriculture" refers to farming where fertilization, weeding, seeding, etc. take place based on the actual condition of each small section of a field, rather than on a whole-field shotgun approach. The actions are tempered by the knowledge of how the land has been used in the past and its planned uses for the future. Some of this input comes from imaging the fields, often using satellites. After these data are analyzed, the "focused agriculture" application instructs the tractor (which carries a GPS receiver and so always knows exactly where it is) as to what type of fertilizer, disinfectant, and seeds to put down at every moment.[25] The result is just the right amount of weed killer and fertilizer, distributed to match the specific conditions of each section of a field. Ongoing experiments indicate that when this technique is used approximately 40% less fertilizer and weed killer is required.

Technology now also has a hand in the type of seed used—genetically engineered seeds can be designed for different sections of a field to further increase yields.

TOO MUCH INFORMATION?   Some people are concerned that with all of these "gadgets" collecting and remembering our information for us, we may lose the ability to do this for ourselves. Where would we then be when the batteries failed? Similarly, some feel (with some justification) that the hundreds of Email messages that many of us receive each day inject unneeded distractions that actually reduce our ability to think and be productive. For example, consider this statement: *"The pace with which we receive information nowadays gives us no time to think."* Because many of us get over 200 Email messages each day, there would seem to be some truth to this.[26]

Now consider this statement was made by Lord Melville in 1860! He was referring to the new railroad mail service that delivered correspondence faster and more frequently! It seems that whatever information overload we seem to be feeling, we do have the capacity to deal with more (good thing!).

20 Miniaturized computer-controlled devices to administer insulin and to monitor the heartbeat are already in use, and specialized "patches" are being used in a Rotterdam hospital to monitor "lazy eye." See www.biokey.com/biokey2/projects.htm Additionally, devices such as an "intelligent scalpel," which will only cut the type of tissue of interest, chips embedded in humans that can monitor glucose levels and administer insulin as needed, and more such devices, are under investigation. See bioinstrumentation.mit.edu/projects.htm

21 Spectrum, May 1996. Bioelectronic systems for visual impairments.

22 See www.wired.com/news/news/email/explode-infobeat/technology/story/20517.html

23 www.compaq.com/rcfoc/19990531.html#Putting_Technology_To

24 www.compaq.com/rcfoc/19990614.html#Convergence_And_Kids

25 Earthwatch (www.digitalglobe.com) provides a photo service that offers 1 m² resolution pictures of any place on Earth. Submit your request via the Internet and they will take a picture, within a few days, to fulfill your order (the result coming back by Email). The cost is less than $600. Could surveyors become an endangered species? Another company, Spaceimage, is now providing one-third meter resolution color photographs—check out spaceimage.com/newsroom/releases/1999/aerial.htm for a look at just how good this is!

26 A 1999 Pitney Bowes study, described in the June 11 Ninfomania NewzFeed, reports that the average U.S. worker receives about 200 Email messages per day! Workers in the U.K. receive about 170 per day!

ON THE ROAD How about an intelligent car? It would be continuously connected with traffic-monitoring centers that would be able to suggest the best route. Such a car would always know the best way to go, taking into account construction, traffic, and even open parking spaces.

Today, cars can easily know where they are through the use of inexpensive GPS receivers with a notebook computer (or some dedicated equivalent) that displays moving maps. Most will also let you type in a destination address and then provide you with the fastest route, including turn-by-turn spoken directions.[27] If you have mobile access to the Web, many large cities have sites providing real-time traffic information that you can now manually, and later automatically, use to guide your route.[28] You can also, already, receive textual traffic updates on a Seiko watch.[29]

A "smart car" would break down less often, because it's constantly monitoring itself for problems and automatically "discussing" them with your chosen service center. In many cases, a problem could be detected before leading to an unsafe and expensive catastrophic failure, and if a major problem were about to occur while you were on the road, you might be instructed to pull over and shut down before you stalled in the roadway (and perhaps before major damage occurred). Of course, your "auto club" would be automatically dispatched to come and retrieve you.

If we take these automotive navigation and automation systems a step farther, will we find that drivers become unnecessary and that we can sit back, chauffeured around by our silicon servants? We're a long way from that, but experiments in self-driving cars have been underway for several years on a specially instrumented section of Interstate in California.[30] Once this does become "real," it might be very nice after some wine with dinner or if you're traveling through a dense fog![31] Another benefit: European studies have indicated that automated highways will be able to handle far more cars per hour than we're used to today.[32] But while we *are* still driving at night, wouldn't it be nice for technology to pierce the darkness?[33]

"Smart" communicating cars can provide other services. Are you locked out? Use your pocket cell phone to call your car's service provider, and with proper ID, they'll let you in.[34] A similar mechanism might be used to deny access to the car (the car might know it's off-limits to your children after a certain hour), or, perhaps, the car dealer might disable the car as a "gentle" reminder that a payment was missed.[35]

27 Example vendors include Delorme (www.delorme.com) and TravRoute (www.travroute.com/).

28 See www.wsdot.wa.gov/PugetSoundTraffic/

29 See abcnews.go.com/sections/tech/DailyNews/traffic960623.html

30 www.compaq.com/rcfoc/970804.html#Driving_the_Information

31 An Italian research project has already demonstrated 1200 miles of autonomous, self-driving by an experimental car (see millemiglia.ce.unipr.it/ARGO/english/desc.html).

32 "Drive": European research projects for transportation beyond the year 2000 (www.trentel.org/transport/research/ISTimages/IST.html#Transport).

33 See www.thecarconnection.com/columns/caught/caught2-1-99.asp

34 The On Star service (www.onstar.com), for example, already detects accidents, breakdowns, and car thefts, and automatically reports them along with the GPS-determined location of the car. The service can also open car doors and start the engine for you if you lose the key.

35 Smart accessories might also save us money—since January 1999, Italian cars equipped with a satellite antitheft device receive a 40% discount on premiums, which pays for the equipment within two years. Expensive cars with high insurance fees recover the investment during the first year.

Smart accessories may help prevent accidents. Biometric identification (finger prints, retinal scans, etc.) might prevent unauthorized use; radar might apply the brakes at the last moment if we don't—both forwards and backwards; cruise control might automatically maintain a set distance behind the car in front, and so on.

ABSURD?   The idea of making all of these things around us "smart" may sometimes sound absurd. Why would we want to make things so complex? The same has been said before. Our grandparents may have felt that spending a great deal of money for a washing machine was "absurd." Today, you might think it "absurd" to have that washing machine discuss the best wash cycle with the garments you toss in to be cleaned. Our kids may not know how to wash a sweater that doesn't instruct the washing machine on how to best, and safely, launder it, or perhaps it won't begin its cycle if it senses something inside that will bleed onto the other clothes. It's all a matter of perspective, biased by the world around us, and our world is certainly changing.

Perhaps you're at work and you've just invited some people over for dinner? Do you have to stop by the store on the way home? Just call the refrigerator to find out what's available for you to cook, and how much is currently left in each package.[36] Did you see an interesting sofa while shopping and wondered how it might look in your living room? Just instruct the store's computer to retrieve the 3D model of your room, and you can experiment with the new couch in context.

Once you do get home, your PDA, which has been constantly updated with your day's information, experiences, and changes to your calendar, might then have a chat with your alarm clock. Because your morning meeting was changed, you get to grab an extra half hour of sleep, and the coffee pot will get updated at the same time so that your morning brew is still fresh at the later hour.

AN OCEAN OF INFORMATION   Whew! Quite a few changes may be in store for us as the things around us begin to think (or at least to store, manipulate, and communicate information without our direct control). Many of these changes hold the potential to make our lives easier and safer; some also have a dark side that we must also come to grips with.[37]

But one way or another, we do have to learn to sail the oceans of information that are surrounding us, without drowning.

36 See www.nikkeibp.asiabiztech.com/Database/98_Apr/30/Mor.04.gwif.html

37 See innovate.bt.com/viewpoints/pearson/yourlife.htm for some twists you may not have considered.

Technologies for Intelligence

Few people find it easy to define what intelligence is, but many agree that intelligent behavior requires three basic components: information processing, storage capabilities, and interconnections among the various elements.

In a biological brain, these three capabilities are provided by structures made up of cells. Whereas, in a computer, these three elements are provided by different components.

Let's take a closer look at these three factors and see just what technology can offer for the future.

CHIPS AHOY   Where the human brain uses cells, most of today's computers are built around transistors. Every year the world's production of transistors[1] (an "information processing molecule," if you will) is about twice the total number of transistors produced to date. For example, 100 million billion transistors[2] were manufactured in 1997 (roughly equivalent to all the ants on Earth); it previously took from 1946 to 1996 to make that many! In 1998, the number of transistors produced increased to 200 million billion, equivalent to 6 billion transistors produced every second.

Not only are there a lot of these little guys around,[3] but they're getting dirt cheap. The price of a transistor in 1959 (when they first entered serious commercial production) was $5; in 1999, that same $5 will buy you 40 million of them for the same price (inside a chip)!

These transistors have done something virtually unheard of in history—the number of transistors on a chip has been consistently doubling every 18 months, although the cost for those twice-as-many transistors remains the same[4] and that's the proverbial "nontrivial" feat. Not only must the transistors get smaller with each generation, but the chip's design gets more complex. Even the wires, or "interconnections" between the chip's elements, have to shrink.

---

1 The invention of the transistor has been compared with that of the wheel and the printing press. Review its history and get a peek at what the future holds at www.lucent.com/ideas/heritage/transistor/index.html

2 To get an idea of just how big the world production of chips is, the powder resulting from the cutting of the silicon wafers on which chips are made is used to make the concrete in skyscraper construction more flexible and elastic. Tons of this powder were used to build the Petronian Towers in Kuala Lumpur, Malaysia, the highest skyscraper ever built—(see *Scientific American,* December 1997, pp. 92–101).

3 See www.intel.com/onedigitalday/explore/intro.htm

4 This is known as Moore's Law, named after Gordon Moore, one of Intel's cofounders, who proposed it back in 1975.

This requires increasingly sophisticated equipment for making and exposing the "masks" used to create the map-like patterns that define each layer of a chip.

Another problem in shrinking these on-chip elements involves heat dissipation[5] (the more transistors crowded together, the more heat is generated in a smaller space; getting rid of it is both an art and a science). As we continue to overcome these problems and produce chips with ever-more, ever-smaller transistors, their capabilities will increase dramatically. The smaller transistors typically consume less energy, and the fact that the chip's elements are physically closer together means that the nasty old speed of light limit[6] is less of a problem.

VISIBLE LIGHT   Today, common CPUs contain around 10 million transistors, but we're run-
IS TOO BIG!   ning into a bit of a problem as we try to define ever-tinier elements on the
                   chips—the wavelength of the visible light used to lay out the pattern of lines
on each layer of the chip is just too large! But fear not, a new technology, called EUV (extreme ultraviolet), based on electromagnetic radiation whose wavelength is shorter than visible light, can be used to construct transistors less than a tenth of a micron[7] across. This and similar technologies could, according to Intel's Andy Grove, lead to a 1 billion transistor chip by 2010 that can execute 100,000 million instructions per second (MIPS)[8]—compare that with today's fast consumer chips, such as the Pentium III, that contain about 9.5 million transistors.[9] Merced, Intel's next generation of chips, will likely contain between 20 and 40 million transistors.[10]

Of course those are just the evolutionary changes we can expect. On the revolutionary front, which could make everything we're used to today seem like yesterday's vacuum tubes, we have experiments in using DNA for computing and memory elements,[11] individual

---

5 KryoTech offers a chip cooling technology that provides low-cost cryogenic cooling, meaning that chip speed can be increased by about 30%. (see www.kryotech.com/techtext.asp).

6 Something we really do have to get around to repealing.

7 In 1999, typical CPUs such as the Pentium III are made using .25 micron technology, whereas the next generation will likely shrink elements to .18 microns. Already, the Mobile Pentium II processor is being produced using .18 micron technology, and contains 27.4 million transistors! Intel, AMD, and Motorola are investing heavily to continue developing this technology (see www.intel.com/pressroom/kits/processors/quickref.htm).

8 These numbers refer to the speed on a single chip, but a computer may contain several chips, and thus operate at a higher aggregate processing speed. For example, in 1999, a 256-Compaq Alpha configuration could deliver 154 billion operations per second! See www.compaq.com/rcfoc/19990726.html#Weve_Only_Just

9 At the "system on a chip" level, complete systems built into a chip already have a much higher number of transistors. The 912, a 20 × 20 mm chip, contains 223 million transistors (www.lsilogic.com/).

10 At the end of 1997, IBM introduced a new way to interconnect the elements on a chip, using copper instead of aluminum for the tiny wires. Because copper conducts electricity better, the wires get thinner, less heat is generated, and we end up able to pack more transistors together—up to 200 million as this technology matures, according to IBM. Similarly, Texas Instruments announced a copper- and xerogel-based interconnection that should provide a ten-fold increase in chip speed, making it possible to pack up to 500 million transistors on the chip (see www.ti.com/corp/docs/pressrel/1997/c97091a.htm). Those transistors are interconnected by microscopic wires with an overall length of 1.5 kilometers, and all within an area the size of a fingernail.

11 See www.eetimes.com/story/technology/advanced/OEG19990709S0016

molecules (in a new science called *moletronics*),[12] and computing at the atomic, or quantum level.[13] These revolutionary possibilities hold the potential to change *all* the rules.

AN ELEPHANT'S MEMORY   The chips we use to store information have demonstrated an amazing and consistent increase in how much they can remember. Today, 128 million bits of information can be stored on a single commodity chip.[14] This increase in capacity has been paced by falling cost. In 1975, 2 kilobytes of memory cost $125, which means that semiconductor memory cost $62,500 per megabyte. By 1995, the cost had dropped to $41 per megabyte. In 1999, you can buy memory on the spot market for under $1 per megabyte. If these trends continue, by 2002 a megabyte of memory may cost as little as three-tenths of a penny!

These are just the *evolutionary* trends—what if you "change the rules," as Intel did with its introduction of Strata Flash memory?[15] In today's commodity memory a bit of information is (conceptually) "remembered" by storing a voltage in a simple circuit, and the value stored in that "bit" is later read back out by measuring the voltage. Again, conceptually, if the voltage read back out is between zero and half it's maximum possible value, the bit would be a zero; if the voltage was above the halfway mark, it would be a one.

The graph represents the growth in storage capacity (red line) and processing capacity (blue line), expressed in terms of chip complexity (number of transistors). The green line indicates the progressive decrease in costs for memory chips.

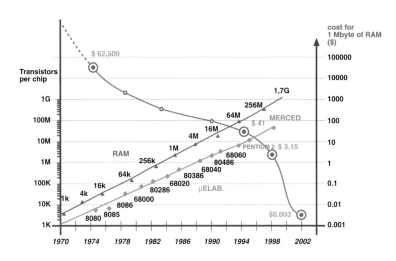

---

12 See www.ortge.ufl.edu/fyi/v26n01/fyi074.html

13 See www.hpl.hp.com/features/stan_williams_interview.html

14 See www.samsung.com/products/128m_flash.html

15 See www.intel.com/pressroom/archive/releases/FL091797.HTM

But chip real estate is a terrible thing to waste, so Intel decided to store and read a finer set of levels in the same circuit, measuring the voltage between zero and one-quarter the possible maximum, between one quarter and a half, etc. This finer discrimination of the voltage allows them to store 2 bits of data in exactly the same real estate where previously they could store only 1 bit! They expect to be able to discriminate even finer steps in the future, allowing 3 bits to be stored in exactly the same space. Moore's Law didn't know what hit it.

Another interesting revolutionary concept being explored makes use of carbon molecules instead of silicon, assembled into buckytubes.[16] This technology may be able to create much smaller circuits (a buckytube is about 100 times smaller than a transistor), but might also be used to create a completely different type of computer system!

The concept is to *not* design logic circuits by arranging the buckytubes in a specific order (the way our chips are designed today), but to develop software to understand the computational characteristics of a *random* distribution of buckytubes and then figure out how to best make use of the circuits that result. Pretty absurd, huh? But that configuration is similar to just how our brains are organized and to how we learn. Don't forget that buckytubes are made of the same basic material as those familiar "carbon-based lifeforms"—you and I.

Remember the alchemists, who dreamed of turning commonly available elements into valuable gold? Well, in a way, we're doing just that today. Consider a pound of common sand, worth about 2 cents. Purify it into polysilicon and it's worth about $38. Melt it into a silicon crystal and slice it into wafers, and the wafers are worth about $1,500. Prepare those wafers for the next step of semiconductor production, and they're worth about $33,500. Now, transform them into chips and that one pound of sand is suddenly worth a half million dollars. Now, sell these chips on the open market, and they're worth an incredible $3 million! From two cents' worth of sand . . .

MASS STORAGE   The cost of magnetic storage, such as disks, has been following a similar higher density/lower cost spiral. In 1984, one of us purchased a "huge" 20 megabyte hard disk drive for a "huge" $1,200—disk space cost $60 per megabyte back then. By 1996, the cost had dropped to 11 cents per megabyte. By 1998 a megabyte of disk storage cost only three cents. In 1999, storage can be had for a penny per megabyte, and if predictions of a 30 gigabyte disk drive for $200 in 2000[17] come to pass, stor-

---

16 This technology is an evolution of the fullerene, a molecule consisting of 60 carbon atoms in a symmetrical arrangement to form a "buckyball." In the buckytube, the structure resembles a tiny tube. It has electrical properties similar to those of a transistor containing a Schottky junction, a point that can be either a conductor or an insulator depending on the applied voltage (see **www.kheper.auz.com/future/nanotech/carbon.htm**).

17 "We're looking at a 30 GB drive by the time the year 2000 rolls around, and for $200." Martin Reynolds, Gartner Group.

age prices will have fallen to an incredible seven-tenths of a cent per megabyte![18] In fact, disk drive storage is actually improving just a bit faster than Moore's Law![19]

Of course you might reasonably be wondering just why we might really *need* the amount of storage these prices imply. Well, as an example, let's consider a "high-end" user—the particle accelerator[20] under construction at CERN (in Geneva) that should begin operating in 2005. It will generate 100,000 billion bytes[21] of information every second it is used. Even with real-time filtering and processing to reduce what has to be stored, they'll still need to find a home for 5 million billion bytes[22] of information—each year.

THE INTERCONNECTIONS   The third element we would like to discuss in this section is the interconnection between computing elements. We already have computers with thousands of chips[23] inside a single box, and this interconnection of many processing units is the best way we have today of significantly increasing computational capacity. If a single chip can process one million instructions a second then adding another allows 2 million instructions to be processed.

Even with that said, things aren't quite that simple, 100 processors don't (usually) get a task done 100 times faster. Although some of the actions executed by a program *can* be performed in parallel, many others must be performed one after the other—sequentially.[24] This limits the tasks that massively parallel computers are (currently) good for even with very special programming.

In addition, where parallel execution does make sense, it requires care to synchronize the different streams of calculations. The bottom line is that a good parallel computer,

---

18 And those are just the *evolutionary* possibilities—various schemes in the labs may make these drives look like floppies. For example, the Feb. 2, 1999 *Nikkei English News* described Fujitsu Research's work on "Quantum-Dot Wavelength-Multiplexed Memory," which could yield 1.1 trillion bits of data stored in something the size of a postage stamp. That's 500 DVDs of data in a square inch, and according to *Upside*, NEC Labs is working on holographic memory that, theoretically, could store 30 terabytes of information in something the size of a sugar cube (www.upside.com/texis/mvm/story?id=3720f0250). Other research is working on using the magnetic field in *single* atoms to record an on-or-off bit. These chips, if successful, might store 1.7 gigabytes in a tiny (18.6 × 7.5 mm) device and be readable ten times faster than current memory chips—www.nanochip.com This kind of changes your ideas of what can and can't be done.

19 See www.compaq.com/rcfoc/19990607.html#Storage and www.compaq.com/rcfoc/19990614.html#And_Denser

20 Large Hadron Collider project (www.cern.ch).

21 100 terabytes.

22 5 petabytes.

23 The Pathforward project has clustered 32,000 Pentium chips into a single structure (www.llnl.gov/asci-pathforward). The project, carried out by the U.S. Department of Energy, has an interconnection system supporting speeds between supercomputers ranging from 2 to 4 billion bytes per second, and processing speeds of around 30 teraflops are expected by 2001 (30,000 billion floating point instructions per second). Today, this speed would require more than 1 million "top of the line" PCs. What's their goal? To reach 100 teraflops by 2004.

24 Latest generation chips already execute instructions in parallel, in terms of operations performed within a single instruction, carrying out parts of the next instruction in advance whenever possible.

combined with a "well-written" program, typically makes use of only about one-third the processors' capabilities. If a massively parallel computer consists of 1,000 processing units, its speed might be equivalent to 300 times the power of a single one. Not bad, but not what you might expect. New software and hardware technologies, though, are constantly being explored to increase this performance ratio.[25]

Who needs this type of computing power? Well, even with today's computers, we still can't accurately model our planet's atmosphere, but that's just what we need to be able to do to better predict the weather. (Two of the authors live in New England, and they can attest that weather forecasts have a long way to go.) Nevertheless, as computing capabilities continue to increase, we will be able to process an ever-greater number of different possible weather outcomes. That will let us pick and choose the results that seem the most correct, and we'll then use those choices to reinforce the next set of choices. But doesn't this process sound familiar? Doesn't this sound like how "we" learn?

These types of "probabilistic" computers, as they evolve, may begin to react in ways similar to behavior that we would consider intelligent. One of the characteristics of intelligent behavior is the ability to be aware ("I'm not just thinking, I know that I'm thinking"). In today's sequential, or even in our massively parallel computers with fixed programming, this "intelligent" capacity has not demonstrated itself. Perhaps, as the number of computing elements increase, and the interconnections between them increase, some thresholds may be passed, and something quite new might take place.

THE INTERNET   The Internet is really just such a vast interconnected network of hundreds of millions of computing elements. Each Internet-connected computer operates on the basis of its local context, but it can have a (limited) influence on what happens globally; for instance, by changing the context in which one or more other computers operate.[26]

Take, for example, Internet discussion or "chat" areas. When there are numerous participants in a discussion, a train of thought, as well as an "ethic," starts to emerge in each discussion area, with progressive clustering of people who "feel" the same way. Even though no messages are censored, the community tends to ostracize people who do not adapt or evolve with the others. A single disturbing element (an idea that goes against the mainstream) may not have any effect at all or may trigger profound changes that are difficult to predict.

---

25 See www.starbridgesystems.com/home/product_comparison.htm

26 Consider the SETI@HOME project, which allows anyone with an Internet-connected PC to run a special screen saver that lets it process radiotelescope information as part of the search for extraterrestrial intelligence. Any one PC is only a minor cog in this process, but thousands of participating PCs around the globe represent a massive supercomputer, the equivalent of 28,000 years of computing time during the first two months that cost the SETI project nothing (see setiathome.ssl.berkeley.edu/).

Today, it's people behind the keyboards who are "chatting," but this example, of negotiating ideas within discussion groups, can easily be extended. Suppose that people were replaced with software programs doing something without direct human control? OK, we know that sounds like science fiction, but that's just what the so-called "intelligent agents" that many companies are working toward may be! In fact, standards (such as XML)[27] and related techniques are now being developed just so that information can be automatically negotiated between automated agents.[28]

---

[27] www.w3.org/XML
[28] See the section entitled "Electronic Commerce," page 105.

4   Landlubbers

In 1998, 1.5 million new pages of information were published on the Internet each day! As of August 1998, this left the Web with 3,000 gigabytes of information. A company that works to archive every page on the Web, Alexa, had 12,000 gigabytes (12 terabytes) of the Web preserved—that's half the amount of information stored in the Library of Congress![1]

Over the last few years the concept of "surfing" has been associated with the use of the Internet, but unlike the fantasy of riding a surfboard toward a beautiful beach, many Internet "navigators" get discouraged as they search for the information they need to solve a particular problem in this vast "information sea."

Surfing the Internet just for curiosity is an exciting and interesting pastime, but if a "surfer" is looking for a specific bit of information to solve a particular problem, things can get much more complicated. In many cases, the information they uncover is not particularly relevant, or they may find so many possible places to look at that it's easy to become overwhelmed. In fact, given the vast amount of information accessible on the Internet and the relative immaturity of the "information tools" currently available, we all are still very much neophytes on the Internet sea.

"I JUST KNEW IT WAS THERE, SOMEWHERE . . ."[2] If you find yourself muttering something like this as you use various Web search engines to hunt for that nugget of information you need, your search technique may not be to blame if your goal remains elusive. According to research conducted by two scientists at NEC Research in 1999, even the best of the bunch of search engines are only cataloging about 16% of the pages on the Web. (Compare that to 1997 when the search engine with the greatest coverage, of an admittedly much smaller Web, covered 34% of pages.)[3]

In 1999, the three search engines with the greatest coverage are Northern Light, Snap, and AltaVista (all between 15.5% and 16%), and

---

1 See www.internetnews.com/IAR/article/0,1087,12_10381,00.html
2 From the "Rapidly Changing Face of Computing"—www.compaq.com/rcfoc/19990719.html#I_Just_KNEW_It
3 According to the July 8, 1999, *New York Times* (www.nytimes.com/library/tech/99/07/circuits/articles/08geek.html), the study's authors feel that, *"Search engines are increasingly falling behind in their efforts to index the Web."*

that's a significant difference from the coverage of the two lowest-ranking engines, which they found canvassed only 2.5% of the Web.

Perhaps this is one reason that Nielsen/NetRatings has found that the growth of traffic to portal sites, many of which are anchored by their search engines, is not growing as fast as the Internet's population, overall.[4] Of course before throwing stones at the search engines, we might want to remember that the Web is a huge place; the study's authors estimate that in 1999 the Web was made up of 2.8 million sites containing 800 million pages and 180 million pictures—a total of an incredible 15 terabytes (15 million megabytes) of data,[5] and that doesn't include the myriad of "pages" of information hidden within databases that only become "pages" to answer your query.

Interestingly, the study also attempted to categorize the pages on the Web, finding that 83% were commercial in nature, 6% related to scientific and educational material, and to the surprise of many, only 1.5% contained adult content.

So what do you do until (if ever) the search engines cover the Web from end to end? The *Times* suggests that if you can't find what you want from a single search engine, you might try one of the "meta" search engines, such as MetaSearch,[6] which dispatches your query to multiple search engines and attempts to consolidate the results.

BOOKS IN THE INTERNET AGE

The Internet produces, consumes, and stores information at a pace unknown in history, and so this "Info-problem" is only going to grow. For example, books are being transformed into bits on the Web. Libraries[7] around the globe have begun to convert miles of bookshelves into gigabytes of bits.[8] Ancient parchments that were once available only to a select few are now becoming accessible to us all on the Web,[9] expanding research opportunities enormously. In many countries, new laws, court decisions, and other legal information are being published on the Web, making them accessible to the masses. Similarly, we can all now easily find out about new scientific discoveries; we can even see images coming from Pathfinder on Mars[10] or from the Hubble telescope in orbit around Earth.[11] A world (and beyond) of information is available on virtually any topic under the sun, from gardening, to driving directions, and maps from here to Grandma's house,[12] and even instructions on how to knit! This vast

4 More users are apparently skipping the searching and heading directly for sites they already know how to find—www.zdnet.com/zdnn/stories/news/0,4586,2288110,00.html

5 See www.zdnet.com/pcweek/stories/news/0,4153,1015425,00.html

6 See www.metasearch.com

7 The California Digital Library is one example of a global project to convert all references in a library into an electronic format (sunsite.berkeley.edu/Catalogs).

8 Although this trend is extremely valuable, we should also be asking ourselves how we're going to preserve this information for the future. For example, after 3000 years we can still read the Rosetta stone, but can you read a document stored on the 8-inch floppy disk that was common 20 years ago? How will future generations be able to read the information that today we're increasingly storing only as bits? Do you expect your great grandchildren to have an application that will be able to read a Microsoft Word document you save today?

9 See the Dead Sea Scrolls at lcweb.loc.gov/exhibits/scrolls/toc.html

10 See powerweb.grc.nasa.gov/pvt/Mars.html

11 See www-b.jpl.nasa.gov/s19/hst.html

12 Lucent Technologies and others offer free mapping services (such as at www.mapsonus.com and www.mapquest.com) that allow you to enter a departure point and destination. The service then displays a map of the best roads to take, even providing turn-by-turn directions with "mini-maps" showing the details at each turn!

storehouse of accessible information also enables a new array of new services.

Many applications are becoming "Web-enabled." For example, if you're writing an essay about Shakespeare, one day your word processor will notice this without explicitly being asked and will open a window providing relevant information. You might see a list of the cities where Shakespeare staged his plays, and perhaps critiques from that time. You might find links to religions and the art of that day, and perhaps that period's latest fashion trends. In effect, we're talking about each of us having an army of "intelligent agents," some of which operate on our behalf automatically like the word processor, and others that we might send off to find the best airfare for a vacation, or the lowest price for office supplies. Our agents could know who we are, our interests, and preferences, and they would be able to communicate over the Internet with Web sites, and even interact with other agents, to do our bidding.

A New Type of Librarian

How will these intelligent agents learn about us? We can, of course, "teach" our agents about our likes and dislikes through filling out forms or programming them in other ways. But perhaps of more interest is the idea that our agents will quietly watch what we do from day to day, learning from our actions and the daily choices we make. Don't forget that as more of the things around us become "smart," such as communicating Personal Digital Assistants, electronic wallets, and eventually smart watches, our agents will be able to observe our behavior throughout the day; not just when we're sitting at our PCs. (Of course there is also a "dark side" to such information gathering—we'll certainly want to secure the profiles our agents develop about us from prying eyes, and we'll also have to be alert to other agents attempting to profile us! Privacy laws will become increasingly important.)

So, here we have all these agents ferreting out some information for us, but suppose one of them comes back with something it thinks is relevant; however, it's in a language we don't know how to read? With the Internet, that's not a problem—numerous services, such as AltaVista,[13] will happily translate between English and a variety of other languages (although in their current state, you'll never mistake machine translation for that of a good human translator); but one day, our agents may be able to call on

---

[13] babelfish.altavista.com/cgi-bin/translate?

increasingly effective translation services that are transparent to us, so that the increasingly multilingual Internet seems to be entirely in "our" language, whatever it may be.[14] Other services may even translate one media into another, for example from voice to written text, or, though much more difficult, we might be able to obtain a textual description of an image. Once software is able to do such "human" tasks as understanding the content of an image, even more interesting services will become available.

Looking for a Roaring Lion

For instance, we could search an online movie museum for all its clips about lions, or about people fleeing from a lion, or clips in which a knight has been changed into a lion by a magic spell. Now each of these searches has an increasing level of complexity. Looking for a lion is the simplest case, although it's not really very easy, because a lion must be detected within a picture containing many other elements, or the picture might present the animal from different angles.

The second type of query requires an understanding of images and of what's happening within each image and in a string of images—the lion may be roaring in the first frame, whereas images of people fleeing may appear only in subsequent ones; the software would have to make the connection. The opposite is also possible, with shots of frightened people followed by the lion. The third query requires an even more complex understanding, that of being able to follow the plot and spot a knight that was previously turned into a lion.

Obviously there's no limit to the level of complexity and difficulty[15]— you could even go as far as saying, "I want to see a clip that makes me laugh." In the future, your agent will be able to look for just this kind of scene, explaining to other agents what it has learned would be funny to you.

Once our computers do become capable of such image recognition, a variety of new services will open up. For example:

- One of the characteristics of a good physician is his or her ability to quickly analyze a patient with a glance (sometimes called a "clinical eye"). Why not let our agent use images of us from handy digital cameras to do similar continuous top-level diagnosis? Why not enhance these capabilities with more detailed biometric data that might be gathered when we touch the water faucets in the bathroom, or from our

---

14 Although the Web began containing primarily English text, by around 2001 English pages are expected to have become a minority as a vast number of other-language pages come into existence.

15 It's interesting to note how, in general, the easier something is for a human, the more difficult it is for a computer—the questions about the lion would be easy for us to answer but are extremely difficult for the computer. On the other hand, making millions of calculations to predict the weather is beyond our capability but quite feasible for a computer. To each his (or her, or its) own strengths.

temperature, taken automatically by an infrared camera? Couldn't an "automated physician" perform some of these complex but rudimentary health analyses to let us know when we should visit a doctor? Note that similar computerized services are already becoming a reality. Some programs are being used to read electrocardiograms and compare them to the patient's previous tests; sometimes this provides a better diagnosis than the average physician! These programs have the potential to become even better than a specialist, in time.[16]

• Monitoring people in areas such as passport lines may become more effective through the use of surveillance systems that can single out the image of each person, and compare it with the images in a file of "suspects".[17] It is interesting to note that computerized image processing can "see" a face under a beard or mustache, behind sunglasses, or under dyed hair, in ways that humans cannot; spies may need much better disguises in the future.

• Such image analysis may also come to play in the retail environment, detecting a customer's reaction to a particular product or display and perhaps offering alternatives until the customer becomes "interested."

IF THE MOUNTAIN CAN'T GO TO MOHAMMED, THEN MOHAMMED GOES TO THE MOUNTAIN

The fact that information is available and can be easily retrieved over the Web is not any guarantee that the information will be well used. On the contrary, as the amount of information increases, more of it will be ignored! But the new services we've been discussing, based on intelligent agents and profiling users, are geared to winnow the information mountain down into a digestible molehill, providing just "the best" information, when and where it is needed.

Consider people working in a particular area of a company; in addition to their personal interests, each of them will also be interested in information related to the area where he or she works and to the job they're doing. As an example, a person working in the design department of a company that produces household appliances might normally want to be informed about the sales trends of her company, market forecasts, new products being announced by competitors, and perhaps advertisements for similar products from competitors around the world. On the other hand, when that person transitions to working on a specific project, say designing a coffee

16 See www.rochester.edu/pr/releases/med/future.htm

17 The "Mugspot" project is part of research conducted by the University of Southern California www.usc.edu/ext-relations/news_service/chronicle_html/1997.10.27.html/'Mugspot'_Can_Find_a_Face.html) and by the University of Bochum, Germany.

grinder, they may then want to focus the information they receive to that particular area, starting from the bean and ending up in the consumer's cup.

Of course such information doesn't have to appear on a traditional PC in response to clicking on a Web page. There are many ways to deliver information. For example, a message can appear on a cell phone during a call, or a voice-enabled agent might call when a particularly important piece of information pops up. Our agent might keep tabs on the arrival time of a plane we have to meet while it monitors the real-time traffic situation along our route; it might then page us when it's time to get in the car and venture forth.

TO EACH HIS OWN ADVERTISING   The mechanisms needed to bring just the right information to interested individuals are still evolving. As we've said, this can be done using a profile defined by the user or developed by an observant agent. At other times, our interests might be discerned from what we're doing at that particular time (for example, when using some Internet search engines, along with the "hits," we sometimes find advertisements associated with what we're searching for).

Sometimes, our queries will be remembered and an information provider might build up a profile of our interests over time, using the profile to tailor each interaction in a more positive manner. Of course such profiling isn't new—children's television programs contain toy-related ads, and in some airports, television screens located near the departure gates have ads that change according to each flight's destination and the primary type of passenger (business or vacation traveler) booked on the flight. Some airport TV systems, such as the one at Washington's Dulles airport, detect when people are near the screen and turn up the volume to attract their attention. In the near future, we'll find systems (already being tested in research labs)[18] that can read a person's expression, adapt the message, and the way the information is presented, based on the viewer's real-time reactions!

THE RIGHT INFORMATION AT THE RIGHT TIME, AT THE RIGHT PLACE   The enormous variety of information now available opens new business opportunities that make it easier for people to find and consume just the "right" informa-

---

18 Studies were carried out at the MIT Media Lab. See vismod.www. media.mit.edu/vismod/demos/ affect/

tion. There are already several specialized companies that, for a modest fee, will filter through large amounts of information and send tailored results to the customer. Today some companies, at a reasonable cost of around $300 a month, will provide you with focused information about competitors, market trends, and technologies that are relevant to your business. Not very long ago such information took much longer to be gathered and cost far more. The difficulty has shifted from the gathering of information to its effective analysis and packaging and use.

Another approach to the successful search for information is the "concentration" of information in places that afford easier access. For example, there are numerous companies now putting millions of job opportunities and job seekers' resumés on the Web, together with tools to match up this supply and demand.[19] Of course this is just the type of job that seems perfect for intelligent agents, and new developments, such as the XML language, which will make it easier for agents to talk together, promise to change tomorrow's job markets.

Overall, as a greater percentage of the world's information does begin to reside on the Web, sometimes exclusively, it will be increasingly difficult for people *without* access to the Internet to be part of the emerging global society.

If we had to lump the changes we've been discussing under one heading, it would be that we're moving from the days of one-way, passive information (newspapers, radio, and TV), to a world of interactivity, where each user can take control as an actor in the world of information. So, let's now take the information stage.

[19] For a site providing pointers to other sites with job offers and guidelines about how to prepare your resumé and get ready for a job interview, see www.smartbiz. com/sbs/careers.htm

Beyond the Physical

At the physical level we have a growing number of ways to gain access to the riches of the Knowledge Age, from the telephone, to the PC, to GPS-equipped cars and pockets,[1] to TVs, and even through some credit card-sized "Personal Digital Assistant" devices.[2] But if we look beyond the actual physical devices, to the "logical level," we actually make use of the Internet through a series of services provided by many vendors. In this chapter we will explore this "logical infrastructure" and how it works.

The traditional telephone system was built around a point-to-point (or peer-to-peer) model. For example, a fax can be sent directly between two compatible fax machines. In contrast, television and radio are based on a broadcast model; a television station broadcasts a program that is simultaneously received by thousands or millions of TV sets.

In a similar vein, the way we access the programs and information through our computers can follow different models. For example, a stand-alone video game or PC contains everything it needs, programs and data, on its game cartridge or hard disk. Alternatively, a networked PC might choose to run an application that resides on another computer, "sucking it in" from that "server" for the duration of the activity. In the same manner, the PC might access data that are stored on a remote server (the "client/server" model of computing.)

For example, we might be watching a program broadcast over TV or via the Internet (broadcast model), and as we watch, we may want to request additional information on the topic (client/server model). Once I've received this additional information, I may choose to interact directly with the person giving the broadcast to steer the course of the talk (peer-to-peer model). The coming together of these typically separate ways to communicate is something very new and powerful.

---

1 *GPS: Global Positioning System.* This device (about the size of a walkman) picks up signals from a series of U.S. government satellites and can pinpoint your position to within 100 ft. Coupled with appropriate software and a notebook computer or dedicated device, it can show your location on even the smallest of back roads, and actually "talk you through" a trip by giving you directions as you approach each turn. It wouldn't surprise us if, in a few years, GPS is provided by default in most new cars—the next "car radio." (**www.navcen.uscg.mil/gps/**).

2 The REX credit card-sized PDA from Franklin is one example—**www.franklin.com/rex/rex3.html**

SURFING THE WEB   Let's look at this in more detail, using the Internet as an example. With a PC connected to the Internet we can transmit packets of data between our PC and any other Internet-enabled device around the globe. But just because the destination PC receives the data packets doesn't mean that it is prepared to do anything with them. It has to be able to offer higher level services. For example, a Web server is additional software that knows what to do with a request for information that we send out from our PC's browser. Or consider a hypothetical higher level service that would allow us to find and purchase a particular picture. This service would have to:

- search out all the potential image providers;

- describe the image we want, define the content, the quality of the image, whether we want to be able to manipulate embedded elements of the image, its format, etc.;

- conduct price negotiations[3] and compare results from different vendors;

- pay for and collect the chosen image; and

- adapt the image to the characteristics of the device we're using (PC, television, cell phone display, printer, etc.).

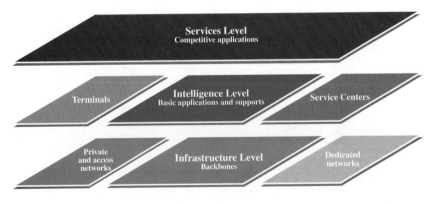

The infrastructure level, along with the access and communication network, provides the physical infrastructure on which information is transmitted. This level supports the network services, forming the intelligence level. This intelligence layer includes terminals and service centers, as well as management applications for routing information, such as those offering "toll free" number services. Above the intelligence level we find the services level. This includes the applications developed by various businesses that compete or cooperate with each other. Some of these applications include electronic mail, information access, job sharing, and virtual travel.

---

3 One of the functions of the MPEG4 standard, described in the section "From Voices, Images, and Words—into Numbers" in Chapter 5, is to carry out negotiations between the requesting customer and possible providers. See www.cselt.it/mpeg/standards/mpeg-4/mpeg-4.htm#E10E14

Today, if we're surfing the Web looking for that "right" image for a report or presentation, we have to perform most of these actions manually, which is a time consuming and complex process. But in the future, as new standards[4] allow our computers to "discuss" things at a higher level without human intervention, activities such as these (and many more) will be streamlined. If you think this is unlikely, let's look into our past to see how this has already happened. In the 1920s a consultant to the telecommunications industry suggested that if telephone traffic were to continue growing at its present rate, by 1970 every person in the United States would have to be working as a telephone operator to manually complete the calls! Happily, that wasn't necessary, because as the call volume increased, the process was automated—and that's just the path we expect higher level services on the Internet to follow. On the other hand, another way to look at what happened with the phone network was that *each* of us became the "operators," remembering those arcane 10 (or more) digit numbers that routed our calls. Hopefully, the growing number of "smart" address books and related services will one day overthrow this trend and, as in the beginning, simply let us say "Central, please call Mom."[5]

The first illustration represents a common pattern of calls made by people in three cities to various destinations using wired, cellular and cordless phones. By applying data mining techniques it is possible to identify certain patterns with unusual characteristics (second illustration). Groups of suspect calls are identified (red and yellow lines) using additional applications that might identify illegal traffic (third illustration).

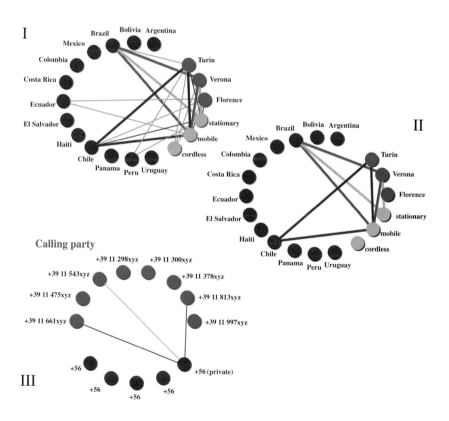

4 Such as XML, or "eXtensible Markup Language."

5 Indeed, we're finding that a lot of today's contemporary telephone services really are "Back To The Future"—see www.cselt.it/Cselt/PUBBLICATIONS/cdig_e/A2.html

CARE FOR SOME HELP? One technique that could help us reach these "easy access to information" goals may be the "intelligent agent." Bringing together software techniques from the world of databases, statistics, artificial intelligence, and object-based architecture, these software programs promise to hide the complexity of all of the Internet's high-level services. In effect, they'll learn about our interests and desires and will do our bidding as if we each had a loyal "personal assistant." For example, if we did have a personal assistant, we might say, "Get me a ticket to Miami for this weekend." Our assistant would know what times we like to travel, how much we're willing to pay, what type of car we want at our destination, as well as the myriad facts about us that such an assistant learns without our ever having to "spell them out." Additionally, our assistant would have a good understanding of the world in which it operates, knowing which airlines are out there, how to access them, how to pay for the ticket, and how to then update our calendar accordingly. We would never have to give our assistant "detailed instructions." These are the (admittedly lofty) goals for the "intelligent agent."

Of course by using the word "intelligent" in relation to these software agents, we're not implying that the software will be "aware" like a human, but rather that it will perform tasks for us, in a way we prefer, that hides the complexity of the environment in which it's working. For example, our intelligent agent might collect and format choices for airline tickets from many sources into one easy to read table, saving us from having to compare many different reports. It might go on to remove entries that it has learned from our prior choices that we don't want (perhaps First Class tickets, "red eye" flights, etc.). As our agents mature, they might go beyond the obvious task at hand. If we ask for an airline ticket from Boston to New York, it might be "smart" enough to realize that what we *really* want is time- and cost-effective "transportation," and it might be "smart enough" to consult train schedules and driving times, presenting us with a comprehensive array of choices we might not have otherwise considered. Now *that* would be a useful agent!

Of course sometimes the information we're interested in can't simply be plucked out of a database, such as an airline's schedule and fares. Suppose, for example, we wanted to instruct our agent to monitor our credit cards and advise our bank, and us, if fraud might be taking place. But what constitutes fraud (assuming our agent can't actually confirm if we're making the charges)? There's no simple definition. But, by using techniques called "data mining," it's possible to figure out patterns from the massive record of all charges and flag those that need to be investigated.

In the future, as our intelligent agents work not only with information that's directly available, but also with information that they can derive from the incredible volume being stored away in "data mines," we'll have fascinating tools to help us improve our work productivity and our recreation time.

5   On Stage

One of the fascinating changes being brought about by the Knowledge Age is the opportunity for people to shed the role of spectator and become participants.

For example, let's imagine being able to receive a football game in a new, specialized, high-definition TV format. It's special camera hangs above the 50-yard line and sends a complete, hemispheric view, which includes the field, the players on the sidelines, and even the fans in the stands—in short, the entire stadium, and in very high definition.

I AM THE DIRECTOR   Because the camera is sending immense amounts of information about its environment, we could, using a remote control and a specialized real-time processor, seem to control the camera in any way we wish. It wouldn't really move—we'd just be decoding and displaying those parts of the common picture that we were interested in seeing. We could zoom, look at specific details—in short, be our own director. We might bring up statistics or other information from a variety of sources on the Web. In fact, if the "licensing" of this special video feed allowed, we could then pass on our own "cut" of the game to others. We really *would* become our own directors!

You see, we've been transformed from spectators into participants.

We've used a football game in this example, but the concept can easily be adapted to many TV shows in which, today, we assume a purely passive role.

This change to active participation has already begun, most notably in video games, some of which are altered by the very actions we take.[1] Another example is the Internet's MUDs, or MultiUser Domains, which can be thought of as vast multiplayer games in which each participant is an actor and affects everyone's environment.[2] Another example is online art galleries that, aside from displaying classical art forms (poetry, music, paintings, or video clips), make it possible to construct your own "original"

1 Broderbund's "Myst" and LucasArts' "Rebel Assault" are examples of games that vary the action, as well as the evolution of the story, on the basis of the player's behavior.
2 MUDs are imaginary worlds in computer databases where people use textual words and language to improvise melodramas, create worlds with all their objects, solve puzzles, and invent games and tools. You assume an alternate identity that, in slang, is called your "character." See www.cwrl.utexas.edu/moo/

art forms. You can, for instance, alter an existing image (respecting copy-rights, of course) in a way that personalizes them; you can even add music that plays when a visitor skims her mouse over the image.[3]

Indeed, many peoples' art can come together into expression that is far more than the sum of its parts.[4]

In the Middle Ages, minstrels crafted a multimedia narrative environ-ment with voice, sounds, and images, constantly evolving the story based on their audiences' reactions. The Web is just expanding the theater from the local town to the global village.[5]

A GIFTED MUSICIAN Another example is the rapidly changing world of music distribution. MP3 is a high-quality music format that most people find indistinguishable from the orig-inal CD, but which produces files small enough that they can be sent over the Internet. Not only can people directly download songs without ever leaving their homes,[6] but if they have (or think they have) musical talent, they can upload their own works to the world![7] This is a far cry from the strictly "top-down," hierarchical music business that is currently reeling under this onslaught. With access to the Internet, for the first time small businesses, even one-person businesses, can compete (to an extent) with the giants.

JOURNALIST AND PUBLISHER Publishing is another example of how the Internet dramatically changes the scene. Consider that just a few years ago it required the resources of a large and probably multinational corporation to research, write, and publish a journal with a global audience.

Yet today, one of your authors does just this from his home office in New Hampshire, using the Internet to research, write, and publish the free "Rapidly Changing Face of Computing" technology journal[8] (which explores the innovations and trends of contemporary computing and the technologies that drive them), and to carry on a weekly dialog with many readers around the world.

One more example of becoming a participant is if you own a Lego Mindstorms® build-a-robot kit. Not only can you go to the Web for Lego's ideas on how to customize your robot, but you can share your own inventions with others around the globe![9] These examples, and the

---

3 Today we have to use a mouse, but tomorrow we'll have far more inter-esting types of "input devices." For example, in late 1999 we should be seeing "force feedback" mice that "feel" textures, and some researchers are already experiment-ing with a "field" in which we can nat-urally move our hands, and their motions will be communicated to the computer without our wearing cum-bersome gloves.

4 *Hamlet on the Holodeck* (ISBN 0262631873), Janet Murray, researcher at the Computing Initiatives Center at MIT, Free Press (web.mit.edu/jhmurray/www/hoh.html).

5 Although very basic, some of today's "electronic books" allow a user to record his or her own annotations, which can then be seen (or not) by subsequent readers of that electronic book. This is being extended to the Web. One company, ThirdVoice (www.thirdvoice.com) is offering software that will allow anyone to annotate any Web site. The annota-tions are kept on ThirdVoice's central server and linked onto the target Web page through special software.

6 See www.mp3.com

7 See www.mp3.com/newartist

8 See www.compaq.com/rcfoc

9 See www.lego.com/webclub/signin.asp

vast number of personal home pages on the Web, and the thousands of "WebCams" which bring pictures of peoples' lives,[10] demonstrate the interest that people have in becoming more active players.

Even for businesses selling physical goods (those made out of atoms rather than bits), the Knowledge Age opens up fantastic opportunities. Suppose we want to purchase a new shirt; it's easy to shop online catalogs, which may have more up-to-date information than a paper catalog mailed weeks earlier. And in the future, the Web may be able to put that shirt onto a (previously scanned) three-dimensional model of our own body to show us how it would actually fit! And remember that "force feedback" mouse we discussed earlier? It might actually allow us to "feel" the texture, softness, and flexibility of the shirt's fabric![11]

MY SIZE, PLEASE!   Of course, if we don't like the look of how the shirt drapes on our 3D model, all is not lost. Because that 3D model is based exactly on our measurements, why couldn't we ask the manufacturer to create a garment that fits us exactly? This is just one more example of how the Internet is an incredible "enabler" for new business ideas.

The list goes on—have you seen the computers at some department store cosmetic counters that capture your face with a digital camera, and then allow you to see how each of their products, or combinations of their products, might look on you? Why not extend this into your home, in the same way that very inexpensive programs already allow you to try on hundreds of different hair styles without cutting a single hair. But suppose while doing this you had a problem finding just the "right look" for you, and you wanted "professional" advice? Why not invite a "consultant" to work collaboratively with you, viewing your results and perhaps modifying them "just so," right over the Internet? (Or, if you turn out to be good at this, why not offer *your* services to others!) This is a good example of how the Knowledge Age, powered by the Internet, opens a vast array of doors. Of course, on the Internet, it doesn't matter if those doors are next doors or 10,000 doors away.

10 See www.comfm.fr/webcam/indexa.html

11 Or, if you don't like the fabric, why not design your own? Scotts House in London already offers its customers the opportunity to design their own tweed by combining different colors of yarn and seeing the result on-screen before placing the order! See www.tc2.com/Home/HomeMass.htm

From Voices, Images, and Words— into Numbers

THE ANSWER TO LIFE,
THE UNIVERSE,
AND EVERYTHING:
IT'S "FORTY-TWO"

Five hundred years before Christ, Pythagoras held the belief that "everything is a number." For example, he discovered that musical notes had a certain relationship with each other; if the length of a string was cut in half, plucking the string would produce the same sound but at a higher octave.

But his ideas didn't catch on. The Greeks had no love of numbers. Their mastery was in manipulating line segments or circumferential arcs; in fact they devised methods (geometry) to perform what were really numeric calculations but without having to use numbers at all. It was only much later, in the seventeenth century with Descartes, Newton, and Leibniz, that geometry problems began to be presented in numeric terms. Even later, in the nineteenth century with Fourier and other mathematicians, mathematical systems were used to represent generic geometric forms, such as an audio waveform, in numeric terms.

Fractal theory has provided us with mathematical expressions that can represent forms found in nature (fractals use small, almost equivalent elements that when infinitely repeated and grouped in various ways, allow for the creation of things as diverse as a snowflake, a tree, clouds, or even a face).

Building on all of these techniques and more, today we can represent many aspects of the world around us with numbers; a good thing since it's really only numbers that our digital computers can directly manipulate.

But there are still many mathematical challenges ahead—we can't yet, for example, mathematically represent emotions, even if fractals do allow us to numerically represent the faces that express them.

THINGS ARE NOT
WHAT THEY SEEM!

Depending on the type of information (if it's text, an image, sound, video, etc.), different methods are used to convert them in numeric form. For example, in the case of western text, each letter is associated with an 8-bit code. According to a convention called ASCII, we say that an "A" is represented by the decimal number 65[1]; if any computer receives the number 65 where it expects

---

1 For example, the letter "A," by the convention known as ASCII, is represented by the decimal value 65, a "B" by 66, etc. See members.tripod.com/~plangford/index.html for the full table.

a text character, it will display the shape for an "A" in whatever font it's using. But the number of variations in an 8-bit code (256) are not sufficient to also represent the characters shape, style, colors, or other attributes (indeed, in some languages, there are more than 256 basic "letters"). So, we have to add more bits. Similarly, we can numerically represent the state of a switch—if it's on, it might be represented as a "1," and if it's off, as a "0." We can use numbers to represent the position of an airplane in relation to the Earth and for so many other things. But there are also many cases where numbers just aren't good enough for representing information.

CODING PHENOMENA IN NUMBERS

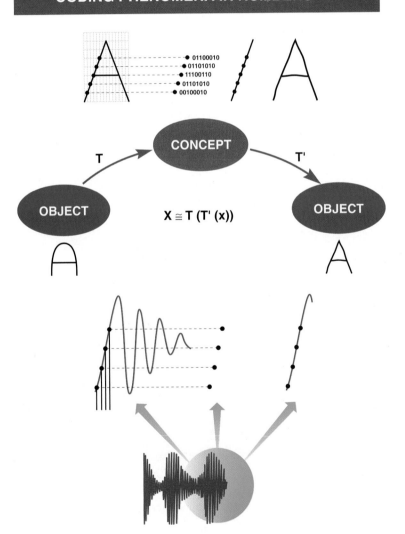

In the first example, which is a letter "A," a grid is superimposed on it and either a 0 or 1 is assigned based on whether or not each point of the "A" appears in that grid square. The resulting zeros and ones are sent to the receiver, where it marks those squares on a similar grid every place it finds a "1." The results is a recognizable "A," but not a very good one (because the grid is fairly large.) In the second example the letter is recognized as an "A," and a commonly accepted value is associated with it. No matter what the form of the letter "A" it is always represented by the same commonly accepted value, and so the "A" can be reproduced in great detail at the receiving end, even though only a small bit of information (the commonly accepted code for an "A") is transmitted. In the third example a sample of sound is taken at regular intervals, and a numeric value is associated with each sample (the "volume" of the signal at that moment). The more often the signal is sampled, the more faithful the digital version will be (at the cost of storing and transmitting more information).

Consider handwritten text, which doesn't lend itself to such coding, because everyone's script is different. Typically, a computer capturing handwriting will acquire and store it as an image, where there is no direct correlation between the script within the image and the ASCII codes of the letters the script represents. Handwriting recognition is getting better, but given that some of your authors have trouble reading their own handwriting, this is a serious challenge for a poor computer.

Back to the image of the handwriting. Images are stored as a series of numbers, each number representing the level of darkness (or color) for a specific location within the image, and the smaller the spots into which we break down the image, the more detailed the computerized image will be (at the cost of taking more time and more storage). This same approach, of breaking information down into an ordered set of numbers, can also be used to digitize voices, sounds, and video.

But the process of digitizing multimedia information, where various media are tied together (as with a TV program, with video and audio both present) adds another dimension. Not only do we need to encode the various types of information, but we have to carefully keep them synchronized! If you've ever noticed a film where the voice seemed to come slightly before or after the lips' movement, it's quickly obvious that something is wrong. But difficult as this is to do "right," and while multimedia is relatively new to computers, people do live and communicate in a multimedia world and it must be done correctly.

Face-to-face communication between people leads to interactions that often unfold in unexpected ways, precisely because of the rich, multisensory environment that encourages you to notice one thing while saying something else, leading to a change in the conversation's direction. So multimedia communications are hardly new. What is new is harnessing technology to provide spontaneous communication, rooted in rich interpersonal relationships, at a distance.

We're already used to multimedia, such as movies and TV; however, they're one-way experiences. Now, the computing and telecommunications revolutions are about to extend this into a two-way, distance-irrelevant experience.

BEHIND THE GLASS TUBE   How does TV bring multimedia into our lives?

Each second of U.S.[2] television video is actually made up of approximately 30 still pictures called "frames," and each frame is made up of 525 "scan lines," with 720 individually colored locations ("dots") possible along each line (you can actually see these horizontal scan lines if you look closely at a TV screen). Each of these dots has a value

---

2 This discussion is about the NTSC standard used in the United States and some other countries. Other systems, such as PAL, which is used in much of Europe, and SECAM used in France, are incompatibly different (don't bring your TV to Europe), but they do operate in a similar manner.

for hue, color intensity and brightness associated with it, which is encoded in 16 bits of information for each dot. In addition, we have sound (not as good as that from a CD), at the cost of another 1.5 million bits per second. All in all, this requires around 167 million bits per second of data for a TV picture![3]

With today's analog modem speeds, even with the newest 56,000 bits per second modems, we would need to increase their speed more than one thousand times in order to transmit the information from a standard TV picture. So how can we hope to transmit full-screen, full-motion video to our computers? We're working in two simultaneous directions: to reduce the amount of data required to represent the content of the TV signal, and to come up with new ways to send the information faster. Let's see how this is accomplished.

We have already mentioned that each second of a television broadcast is composed of 30 still-picture frames. If we convert each frame's image into a series of numbers (digitize it), we can use the massive computing power that we all now take for granted to do some interesting things to it, such as taking out excess or redundant information, thereby reducing the amount of data that has to be transmitted!

For example, if we notice that 20 dots along one line of the picture are exactly the same, instead of transmitting 16 bits of information for each of those dots, we might transmit the information from only the first dot, plus a short code that indicates that the next 19 dots are the same. The "receiving" software on the other end does the reverse, noticing the special code and inserting the "missing" 19 dots where they belong. The actual compression algorithms used are far more complex, but this gives you the idea of how digital compression, which you may already be familiar with from PC ".ZIP" files, works. What we've done is trade off the use of computing power, which we have plenty of, for the (currently) scarce commodity of bandwidth.

Another form of compression for moving pictures (sometimes combined with techniques like those above) involves sending the first frame of a string of frames of video in its entirety. But before subsequent frames are sent, they're analyzed to see how they're different from the first, or "reference" frame. If the video is of a car race and there's lots of movement, virtually each frame may be different and it, too, will have to be sent in its entirety. But if the picture were of a static slide, very little would be different, and over the 30 frames in a given second, far less data would have to be sent to the receiving end (which reconstructs the missing or partial frames based on control information that describes how the encoding was done). Periodically, a new "reference frame" is sent in its entirety, and the process begins again.

---

3 If you do all that multiplication you'll find that it doesn't equal 167—the difference is caused by some "lines" that aren't used for the picture information (which include the "VBI," or Vertical Blanking Interval, where some data are broadcast for special purposes).

Typical video, of course, falls somewhere between these two extremes, and the technique can save a reasonable amount of data. Of course, there are other techniques in use, and even newer ones on the drawing boards that will make use of the continuously increasing computing power that Moore's Law keeps providing.

THE "M" WORD   Moving from the theoretical to the real, in 1989 a group of imaging and data processing experts got together with the goal of creating a standard for digitizing and compressing multimedia information for use by computers; thus MPEG-1 was born (Moving Pictures Experts Group, version 1).[4] This technique enabled video to be digitized into a comparatively small data stream of only 1.5 megabits per second (about 100 times less than uncompressed digitized video). One trade-off was quality; MPEG-1 video was comparable to that of a standard VHS VCR. But that was "good enough"; MPEG-1 became the standard that fired up multimedia CD-ROMS, and it also enabled the first practical uses of (small pieces of) video on the Internet. Of course, putting this in perspective, MPEG-1 video would fill up a floppy disk each second, and this 1.5 megabits per second is still far faster than a typical modem. But, of course, this was just the beginning.

In 1992, MPEG-2 came out, generalizing the standard so that it could work over a broad spectrum of media (CD-ROMs, satellite links, etc.), and allowing various quality trade-offs, letting the content developer balance quality against the disk space and bandwidth available. Of course, there's no such thing as a free lunch—quality high enough to satisfy broadcast applications still requires 4 to 6 megabits per second.[5]

Both MPEG-1 and MPEG-2 encoding are based on breaking an image up into tiny squares for digitizing and compression, but neither format knows anything about an image's composition. Consider an artist painting a dog over a background. It all becomes one painting—the dog loses its identity, or "objectness." To combat this, a concerted effort by TV, computer, and telecommunications companies began work in 1997 on a new standard, MPEG-4,[6] which begins its encoding with any objects that are defined by the director or the artist who created the image. Those objects retain their identity even after the image is decoded.[7] This has a great deal of potential value.

---

4 For more in-depth information on MPEG technology, see www.cselt.it/mpeg and www.mpeg.org/MPEG/

5 But don't scoff at the ability to send broadcast-quality video over "only" a 6 megabit per second pipe. As bandwidth becomes more available, these techniques are going to change how we do a lot of things—like what we see at the movies! Rumor has it that the next Star Wars movie will be shot entirely on digital video tape. It will then be distributed to selected theaters directly over high-speed digital links, where it will be displayed by the seriously big brother of the LCD projectors you may use in your office to display a notebook computer's slide show. This would be the first "filmless" major commercial film. Come to think of it, here we go again—we suspect that we'll still be calling them "films" long after celluloid languishes in the vaults; just as we still "dial" a phone.

6 We know—MPEG-3 is missing. It's a mystery; actually, it was an HDTV-related compression scheme that was rolled into MPEG-2.

7 This objectness is something you may well have already experienced if you've used certain types of computer drawing programs. In some "bitmap" programs, it's like a painting—as soon as you draw a line, it becomes an inexorable part of the picture. Other vector-based or "object-oriented" programs, treat every line and every shape as an individual object, allowing you to move them around at any time.

A NEW AGE FOR IMAGES   Once an image or video stream contains individual objects, both the computer and the viewer can interact with them. For example, you might cause a car to disappear from a scene so you can see what's behind it. You might be able to enlarge a particular object in the scene, like a dog, to study it better. Any given object might be turned into a hyperlink, allowing you to click on, for example, a scrumptious-looking entree being carried through a restaurant by a waiter (regardless of the fact that the dish is moving across the screen), and be taken to its recipe Web site.

But beyond new user interaction possibilities, "objects" hold the potential for using less bandwidth. For example, picture a scene with a house in the background and a couple of people walking in front of it. Traditionally, a full picture has to be continuously transmitted. But with MPEG-4, you might only have to send the house and background once, and then have the computer move the objects (the people) in front of it. Additionally, through the mathematical magic of fractals, many objects—perhaps even faces—might be represented by very svelte formulas instead of large "pictures."

Additionally, because objects are mathematical entities which can be manipulated by the computer that receives them, they can do things you might not expect. Consider the mathematical model of a face—as part of a research project at Compaq's Cambridge Research Lab, one of your authors' faces was digitized and converted into a mathematical model. Then, the text of each issue of the weekly technology journal he writes[8] was fed into a program that not only turned it into speech, but animated the "face" so "he" appeared to be actually speaking the words. The mouth, eyes, and facial muscles moved with the words in a lifelike manner,[9] and even emotions, such as anger and happiness, could be displayed.

PROTECTION   Of course, this ability to digitally manipulate images, video, and sound comes with a "dark side." It's now easy to make unauthorized combinations, derivations, or copies of copyrighted or otherwise protected works. How can an artist even prove that a "knock-off" of her work is in fact based on her intellectual property? One answer is "digital watermarking" technology. Just as a translucent identification word or mark is often imprinted in paper to prove its authenticity, MPEG-4 allows for similar identification information to be subtly encoded into a multimedia data stream. As with a paper watermark, it's invisible under normal circumstances and it can't be removed without damaging the original. But unlike its paper counterpart, a digital watermark is replicated whenever the digital file is copied, and a special program can examine an image, video clip, or sound to see if it contains identifying watermarks, and if so, to whom they belong. This won't prevent copying, but it will

---

8 The *"Rapidly Changing Face of Computing,"* www.compaq.com/rcfoc/
9 See www.crl.research.digital.com/projects/facial/history/history.htm for additional information on "DECface" (and no, that particular face is not your author's!).

help to trace the legitimate ownership of ethereal digital content, and to defend that ownership if necessary.

TOMORROW   The MPEG trail doesn't end there. MPEG-7 goes a step farther along the "objectness" road. Instead of the content author having to identify each object, MPEG-7, being researched currently, will automatically detect objects unaided (opening up the vast libraries of existing multimedia content to benefit from the object technologies we've been discussing). It's not just the ability of MPEG-7 to identify objects that's interesting—if it can, for example, automatically identify every car in a movie, then that suddenly allows us to make queries about things in it and other movies! Imagine going to a video rental Web site and asking, "List the movies that have 1991 Corvettes in them"; or, "Give me a summary of the movies that show the Grand Canyon"; in effect, intelligent information searches of multimedia files.[10]

And as information is becoming both a valuable commodity and a competitive advantage, advances such as these can change the world.

---

[10] The idea of computers automatically recognizing things like cars in movies is not as farfetched as you might think. For example, a growing number of countries are now using traffic cameras that use "computer vision" techniques to not just take a picture of a car that's speeding or running a red light, but to actually interpret the image and extract the numbers of the license plate—and then send the owner a ticket! See www.compaq.com/rcfoc/19990816.html#_Toc459030169

6   My Next Door Neighbor Lives 10,000 Doors Away

ALL ALONE?   We've all heard this cry of alarm: "Computers and the Internet will turn us into people with no relationships! We'll live shut up in a virtual world and never meet and interact face to face!"[1]

It is true that today many people spend less time in direct interaction with others than in the past. Then, the only form of entertainment and information exchange was through face-to-face contact in the town square or the pub, but that's now often replaced by the TV[2]—yet the TV is a one-way, "receive-only" medium.

The Town Square of the Knowledge Age

The Internet, however, allows us to open this window outside our homes bi-directionally. Because the time we spend on the Internet tends to replace some otherwise passive TV time, it actually increases the time we interact with others, albeit in ways and using tools that are different from the glass of beer and darts in the local pub.

NEWSGROUPS   For example, one of the first such interactive environments was the Internet Newsgroup, a text-based "thread" of discussions that predated the World Wide Web[3]; people would see each message and their responses in a "chain," adding their own as they wished.[4] There are thousands of formal Newsgroups covering almost any subject you can imagine,[5] and anyone can create a new group—all it takes is 100 other people who are interested in the subject and who show interest in your proposal for the new Newsgroup.[6]

Interested in Martians, Growing Orchids, or Dog Names?

No matter how bizarre or narrowly focused your proposal, it shouldn't be hard to collect 100 interested people from the Internet's 179 million.[7] Then, everyone can participate. Of course that can be a double-edged

---

1 See www.apa.org/journals/amp/amp5391017.html

2 About 4 hours each day!

3 Although these Newsgroups are now available through the Web—www.deja.com/info/idg.shtml

4 Newsgroups operate in a "store and forward" manner; entries from individual are consolidated together and sent around from computer to computer, assuring that everyone will, eventually, see the extended conversation.

5 As an indication, in August 1999 there were more than 40,000 Newsgroups in existence. See www.deja.com/corp/about.shtml

6 Advice on the creation of new discussion groups is available at URL: www.faqs.org/faqs/usenet/creating-newsgroups/naming/part1

7 As of June 1999, www.nua.net/surveys/how_many_online/index.html

sword—some may contribute in a positive way, some may be skeptics, and some may simply wish to disrupt the discussion. By their very nature, Newsgroups are dynamic; they tend to assume their own characteristics, perhaps quite different from the original intent. They may spin off splinter groups as participants identify other common interests. In fact, this is akin to the ebb and flow of conversation at a party, but this party spans the globe. Although Newsgroups are still primarily text oriented, as bandwidth increases, voice and video are likely to follow, improving on the somewhat stilted meetings conducted using today's high-end business videoconferencing systems.

CHAT ROOMS   The "read it and answer whenever you want" nature of Newsgroups,[8] which enabled them to work before the Internet, precludes real-time interaction. Therefore, a different mechanism has now come to life—the chat room. These are virtual spaces that, like Newsgroups, can be opened by anybody who declares an interest. Once you've opened one, you just wait until someone shows up (most chat room services maintain a directory that lists the "rooms," their subject matter, and their participants; this is how people notice and choose to join a chat room). Once people join, whatever they type is reproduced on each participant's screen in approximate real-time, so that everyone can follow the conversation and type in their 2 cents' worth.

Two common implementations of chat rooms are the Internet-based ICQ[9] and America Online's Instant Messenger.[10] In July 1999, AOL reported 750 million Instant Messenger and ICQ messages per day! More people than live in Fort Lauderdale—150,000 people—are newly registering to use these services each day![11] By comparison, the U.S. Post Office handles a mere 500,000 letters per day.

Even more than Newsgroups, chat rooms are very dynamic. Some of the participants might choose to create their own chat room, which can be "public" or "private," at a moment's notice. Although most chat rooms have been limited to text, voice and video are sure to follow. Indeed, as this book was being finished, Excite introduced a voice-chat feature that uses automatically downloaded software in your browser.[12] And prototype 3D virtual environments already abound.[13]

There is one particularly interesting difference between the Newsgroups and chat rooms. Newsgroups, by the nature of their store and for-

---

8 Newsgroups operate in a "store and forward" manner; entries from individuals are consolidated together and sent around from computer to computer, assuring that everyone will, eventually, see the extended conversation.

9 See www.forbes.com/asap/html/99/ 0618/side.htm

10 See www.aol.com/aim/imreg_ template?template= whatis&pageset=aol&promo= 73013

11 See www-db.aol.com/corp/news/ press/view?release=5681

12 See www.excite.com/ communities/chat/voicechat

13 See www.worlds.com/

ward operation, retain a history of everything ever said. In fact, there are stories of people inadvertently posting sensitive or embarrassing material into a Newsgroup, only to find that there's no foolproof way of removing their submission! Chat rooms, on the other hand, do not normally retain a record of the conversations (but note, just because that isn't normally the case, *anything* you say or do on the Internet may be retained, intentionally or otherwise, by any number of participants or intermediate systems along the way).

VIRTUAL COMMUNITIES   However these services are implemented, what we really are talking about are "communities," where people share information, ideas, and emotions.[14] But instead of being limited to the local bar, these Internet-based "virtual communities" can span continents. For example, a town government might maintain a discussion space where the minutes of meetings are posted, and citizens, even if they're traveling, can follow the activity and post their comments and questions. People with specific diseases can share information on the latest treatments. Indeed, members of some such virtual communities sometimes find themselves better informed than their doctors!

Sometimes a virtual community is born from a single person's experience, for instance, Lorna Wendt,[15] who decided to share her experience of marriage and divorce with interested people around the world. In a short time, a core group was formed that discussed aspects of love and economics during and after marriage. The group went on to offer volunteer psychological and legal support to others in similar situations.

Businesses, too, are making increasing use of virtual communities. Especially where members of a team are dispersed around a country or across the globe, the tools and techniques that are used on the Internet for social interaction can also empower business. The problems of differing time zones melt away when Email and discussion groups allow every member to participate fully but on their own time schedule. Indeed, embracing the differing time zones, some businesses establish three teams, each about eight time zones apart, which share a project and work on it, one team after another, throughout a 24-hour day! A team in the United Kingdom might package up their work and send it on to the team in Los Angeles before going home in the evening; the Los Angeles team will then review the U.K. team's work and add their own and then send it on to a

14 See www.ivillage.com/
15 Lorna was the wife of General Electric top manager Gary Wendt (www.equalityinmarriage.com).

16 See www.compaq.com/rcfoc/ 19990712.html#From_Out_of_the

17 The term "avatar" derives from the Sanskrit and signifies descent or divine incarnation into the world. In the virtual world, an avatar is the manifestation of a real person transported into cyberspace.

18 If you'd like a firsthand example of how a world full of "intelligent" avatars might act, check out a couple of games from Cavedog Entertainment: "Total Annihilation," and "Total Annihilation: Kingdoms." Both demonstrate these concepts very well, the first using a mechanized robotic military theme, and the second set in a medieval magical environment (www.cavedog.com).

19 Indeed, the U.S. Department of Defense thinks that watching the work of avatars might be a very good idea—they have commissioned a virtual reality system where enemy headquarters can be reproduced and populated with enemy and "friendly" avatars, each programmed to act as they might in real life. In this way they can try out different strategies and see how they might turn out. Indeed, the actual "friendly" soldiers, who might be tasked to invade the enemy headquarters, might spend some time in the simulation so they're not surprised when they enter the "real" environment!

In another development, a military simulation "game" was being prepared to be "played" in Australia, so of course the ever-present kangaroos had to be included so things looked "real" to the helicopter pilots. Well, to create the kangaroos, it seems that the programmers took a shortcut and simply changed the look of infantrymen and gave them the ability to hop. Now, roving bands of kangaroos were just too much for the helicopter pilots, who swooped down and "buzzed" the herds, watching the 'roos scatter for cover. But they were certainly surprised when, from where they were hiding, the 'roos pulled out Stinger shoulder-fired missiles and shot down the helicopters! It seems the programmers forgot to remove the infantry's weapons.

team in Japan for the next "shift's" work. Finally, at the end of Japan's business day, they'll send the current state of the project on to the U.K. team for them to begin again, fresh from a good night's sleep. It's interesting to note that such a process can yield greater benefits than just the "24-hour workday"—the necessity for each team to review the previous "shift's" work, three times each day, provides a level of "checking" that is often overlooked![16]

VIRTUAL "YOU" In many ways, computer games have pushed the frontiers of business computing. Audio, 3D graphics, inexpensive hardware graphics acceleration, and stunning artificial worlds that game makers pack into a CD-ROM have all led to advances later used in business. Another game construct, that of the "avatar," may make that transition from the game room to the boardroom in the future.

An avatar is basically a stand-in for "you"[17] in a virtual environment. In a fantasy game, you might appear as a mage or an adventurer as you control your avatar through the game. In a virtual business meeting, although there are times when the powers of a mage or the strength of an adventurer might indeed be desirable, you might settle for an avatar that actually looks like you to add an element of "personality" to your online dealings with colleagues and customers. If we combine the idea of an avatar with "Intelligent Agents," might you imagine a future Internet inhabited by billions of avatars, some of whom look like you, interacting with avatars that look like other people, as they carry out our instructions? Like a computer game,[18] imagine being able to peer into this insubstantial world and watch what our alter egos are up to—it might be the greatest show on earth![19]

Of course, we have to be careful—as we integrate these various forms of virtual communications and communities into our personal and business lives, it's becoming increasingly important that we be sure just who we're interacting with and that the information we send and receive is trustworthy.

Access Infrastructures

Our "backbone" communication infrastructure is evolving very rapidly. Although most of us still don't have very fast, or "broadband," Internet access to our homes and small businesses, competition, evolving standards, and ever-better technology are constantly bringing that day closer at-hand.

A WIRE TO EVERY HOME   Twisted copper pair,[1] since the beginning of the telephone era, has been the common way to tie the phone company to its subscribers. In the beginning, it even tied central offices together across continents! More

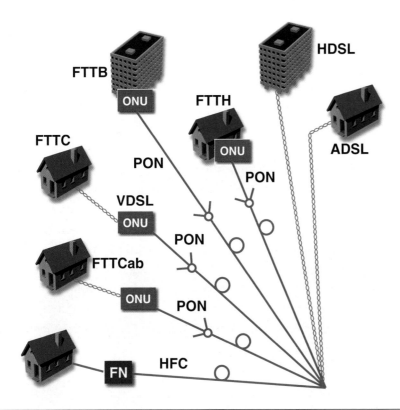

The figure presents different alternatives for high-speed access. These are the primary classes of systems represented: the hybrid fiber-coaxial systems (HFC); fiber-optics systems that can reach directly inside houses (FTT Home and FTT Building) or which stop in close proximity to them (FTT Curb) or else terminate in the closet (FTT Cabinet); and the high-speed transmission systems on twisted pair (ADSL, HDSL, VDSL). The fiber can be split forming a passive optical network (PON). Each fiber terminates with an optical network unit (ONU).

---

1 These are the thin copper wires that go from the telephone company central offices to our homes and businesses. The "twisted" in their name comes from the fact that the individual wires are twisted together to reduce electrical interference when many of them are bundled together. There are a lot of twisted pair wires: more than 100 million in the United States and, at the end of 1997, 800 million worldwide.

recently, the "backbone" linking the central offices has replaced twisted pair wire with, first, coaxial cable, and more recently, with much faster and higher capacity fiber-optic cables.

But change won't end there. In some places wireless (cellular) phone systems are being installed where the cost of a wired infrastructure is prohibitive, and in other areas, enhanced cable TV systems are carrying not only TV and radio but also internet and phone connections. If research by MediaFusion[2] come to fruition, very high speed data might even come right out of the common power plug!

But why this multiheaded evolution away from copper twisted pairs?

What a Twisted Web We Weave

A major problem with twisted pair wiring is that the longer the run is, the less of the signal gets to the other end; the conversation you're listening to becomes weaker. Although today's "active phones" can easily incorporate an amplifier and raise the volume, they also raise the noise the long lines acquire; so there are limitations, especially for data.

Of course, we keep learning to live with such limitations and, to a significant extent, to "best" them. For example, approximately 25 years ago the best modems could only push data across the best phone line at 110 to 300 bits per second, a laughable rate by today's standards. But although the twisted pair infrastructure has remained the same (and actually aged), innovative people have figured out how to use increasingly available processing power to boost dial-up data to 1200, 2400, 4800, 9600, 14,400, 28,800, and finally 53,000 bits per second.[3]

If we move to a fully digital connection [without using a modem (MOdulator/DEModulator) to convert the digital 1's and 0's into audio tones for their trip across the phone network], we can use *ISDN* (*Integrated Services Digital Network*) to get up to 128,000 bits per second over a common configuration. And if we look at the newest "last mile" technology now gaining in popularity, *DSL* (*Digital Subscriber Line*) and its variants, then speeds of megabits per second are possible. Another technology, cable TV Internet service, can offer 10 megabits per second service today (although your PC is unlikely to be able to use but a fraction of that overall capacity.) As they say, ". . . we've come a long way, baby!" But we have farther to go, because, for example, none of these implementations have the capacity to carry individual high-quality video streams to every client at the same time.

---

2 See **www.mediafusionllc.net/northamerica/main/home.html**

3 That last figure, of 53,000 bits per second, is not a typo. Even though you hear of today's modems as 56K, or 56,000 bits per second modems, certain limitations restrict them to a maximum (which you'll rarely see, anyway) of 53,000 bits per second.

LIGHT AT THE END OF THE FIBER   Glass fiber is amazing stuff. Light goes in at one end and, with a bit of attenuation, comes out the other. And fiber can carry far more data than an equivalent length bit of wire. Also, when you bundle a bunch of wires together the different wires' signals can interfere with each other,[4] but by contrast fibers are very "closed-mouth"—nothing escapes to interfere with adjacent fibers (or to be "snooped" by an eavesdropper.) It turns out that mice and other rodents are far less enamored of fiber bundles than they are of our old copper standbys; there are fewer gnawed fiber cables! So, why not replace all the copper twisted pairs with fiber; we'd all have virtually limitless bandwidth!

The answer, of course, is cost. First and foremost, replacing existing cable with new fiber cables is extraordinarily expensive and disruptive—not so much because of the cost of the

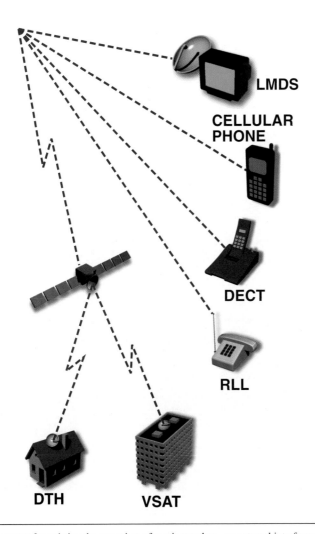

LMDS

CELLULAR PHONE

DECT

RLL

DTH   **VSAT**

The numerous options for wireless (radio) access shown in this figure represent alternatives to wired networks for both telephone services, television broadcast services, and Internet access. The possible choices include narrow band radio access technologies (RLL) for the rapid and economical implementation of telephone services in developing countries, cellular systems, cordless systems, satellite solutions for digital TV, and LMDS technologies for offering wide band services through bidirectional connections.

---

4 That's one reason for twisting the two wires of a pair together—so external interference is minimized.

fiber, but because of the labor involved. Even using new types of radar to look beneath the ground for obstructions, and remote-controlled "moles" to pull new cable, this is your classic "nontrivial task." There's also the matter of the optoelectronics needed at each end of the fiber to convert electrons into photons and back again—they're still expensive for deployment to every home and office.[5] Of course, the cost for all of these components continues to drop every year, so, in the very long term, we may take "fiber to the curb" for granted.

BUT DON'T GIVE UP ON COPPER Copper-based access to the telephone network is still dominant today, and innovations in what this existing copper wire can do are elevating its status. Specifically, *DSL* (*Digital Subscriber Line*) technology is working to change the copper access rules, enabling existing twisted pair wires to carry data at up to 8 million bits per second![6]

This magic is accomplished by sending 256 signals simultaneously over the wire, each at a different frequency (ranging from 4 kilohertz to 1 megahertz), with each signal carrying part of the information. Fast, specialized processors, brought to us by Moore's Law, enable these ADSL modems to perform the complex math required to separate the signal on the transmitting end and to reconstruct the data stream at the receiver.

DSL is still an emerging technology, even as some forward-looking phone companies are beginning to implement it to keep the cable TV and wireless folks from taking over the broadband access pipes.

SPEAKING OF CABLE TV Indeed, "competition" is a major reason why we have so many techniques on the horizon for high-speed Internet access, and in many cities, cable TV is giving the traditional phone companies a run for their money (and their customers). Although it requires a significant upgrade to most cable TV systems, once modernized, they have the ability to provide as much as 10 megabits per second of data to subscribers![7] In fact, with this much data available over a cable TV connection, why couldn't the cable operator also provide traditional telephone service (voice is, after all, just more data)? Indeed they can, bypassing the "Baby Bells" for the first time. No wonder the phone companies are working hard to deploy DSL!

---

5 Lucent, however, has a novel view toward halving the cost by not putting a laser at the consumer end of the fiber. Instead, they interface to the fiber with a nanochip that has many tiny, moveable, controllable mirrors on its surface. The laser from the central office carries information to the home as before, but now, the tiny mirrors move quickly to modulate the reflected light going back to the central office, encoding the "upstream" information from the home on this reflected light! Already, in the labs, the mirrors can move fast enough to deliver upstream data at 10 megabits per second—as fast as a standard Ethernet. Something for (almost) nothing!

6 But there are significant limitations as to the condition of the line and its length; the longer the line between the central office and you, the slower the data. Additional information on DSL is available at **www.adsl.com/adsl_forum.html**

7 This is not as clear cut as it seems, because, unlike DSL, where each subscriber has his or her own chunk of bandwidth, cable TV users in a neighborhood share the 10 megabits per second of bandwidth. So, if lots of kids are at home in the evening surfing the net, things may slow down for each user. This increased neighborhood traffic would not directly affect a DSL user.

With all of these media—TV, telephone, and Internet access—coming over one cable, there are some fascinating potentials for *integrating* the media together. For example, we may see football statistics delivered on-demand over the Web appearing in a separate window as we watch a TV football game, or we may see a product we're interested in on the TV screen and click on it to get a detailed description (including an Order button, of course).

INTERNET IN THE SKY! Between copper wires and now glass fibers, we've come to expect that most of our telecommunications will be done tethered to a phone or network jack. But according to Nicholas Negraponte, director of MIT's famed Media Lab, our future phone calls and network connections will likely be wireless! That would also spell a dramatic change in how our digital devices reach out and touch a world of information.

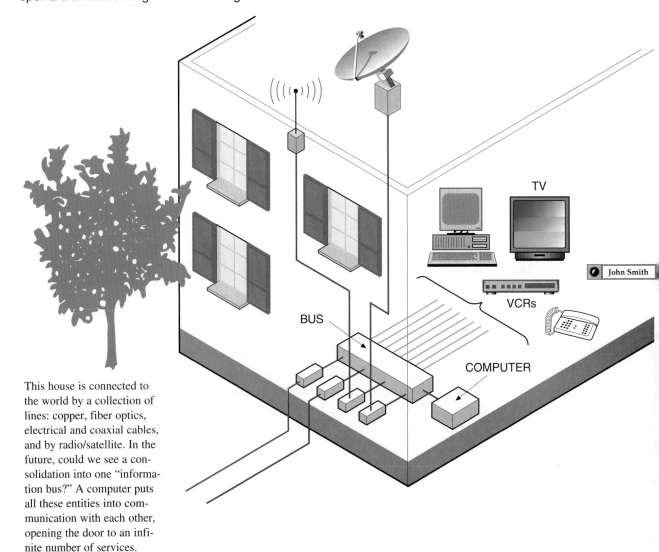

This house is connected to the world by a collection of lines: copper, fiber optics, electrical and coaxial cables, and by radio/satellite. In the future, could we see a consolidation into one "information bus?" A computer puts all these entities into communication with each other, opening the door to an infinite number of services.

Mobility is an important attribute of the Knowledge Age. The pace of competition often demands that people not be tied to a desk or even to an office, but they must have instant access to their business information wherever and whenever necessary. Even on a personal basis, instant access to information can be of great benefit—have you ever been in an airport when your flight was cancelled? Instant access to an airline reservation system might have you booked on the next flight out while the other hapless passengers-to-be are just getting into their long line behind the counter! For these and far more reasons, many companies are working to cut the wired umbilical cord and set us, and our information, free.

For example, on the "office" side, consider the "T1" lines that larger businesses have been purchasing from traditional phone companies for years, delivering about 1.5 megabits per second of data (usually, now, for an Internet connection). Installing such a line often takes months, and the service can cost thousands of dollars per month. By contrast, Innetix[8] is one company that offers an interesting alternative—T1-speed service *without any wires* in the San Jose, CA area for $649 per month (using special radio transmitters on buildings and towers). Another example of a company offering a no-wires replacement for previously wired data services is Warp Drive Network.[9]

But not all wireless data have to come from a ground-based transmitter! One unique solution is being pursued by Angel's HALO (High-Altitude, Long Operation) aircraft; a tag team of three such aircraft would orbit higher than 52,000 feet above a major city, providing broadband "Internet In The Sky" throughout a 50–75-mile diameter area.[10]

Similarly, SkyStation International plans to offer a "Stratospheric Telecommunications Service" from floating blimp-like communications platforms almost 100,000 feet high in the stratosphere.[11]

Then there are the systems that are planning to offer broadband "Internet *Beyond* The Sky." Perhaps the best known is Bill Gates' and Craig McCaw's Teledesic, which, by 2003, plans to place 288 Low Earth Orbit Satellites (LEOS) in a series of polar orbits, such that at least one can "see" every spot on the earth's surface at any time. Initial expectations are that Teledesic will provide up to 64 megabits per second download and 2 megabits per second upload speeds to those who want it. They say their rates will be ". . . comparable to those of future urban wireline services for broadband service."[12] Another satellite-based service, Spaceway, is scheduled to become available a year earlier, offering broadband download speeds and upload speeds as high as 6 megabits per second, although in more limited coverage areas. Of course there are other contenders, er, in the wings, as well.

---

8 See www.innetix.com/

9 See webvn7.webventure.com/warpdrive/

10 See www.angelcorp.com

11 See www.skystation.com

12 See www.teledesic.com/overview/fastfact.html (as of August 1999) for the quote, and www.teledesic.com for more information.

7   Who Said That?

Centuries ago, when barter was the most common form of commerce and the buyer and seller looked each other in the eye before concluding a transaction, there was rarely a question as to who you were doing business with or in verifying the quality of what you were buying—you could examine and touch the goods.

Money changed that; first of all you had to be sure that the currency you were accepting wasn't counterfeit, and if you were purchasing "remote" goods, you could no longer verify their quality. To address this, in the 1300s, the Florentines developed a system of brokers and guarantors[1]—people who ensured that monies would be paid and quality goods would be delivered, using special certificates that they dispensed from their counters[2] surrounding the marketplace.

Seven hundred years ago brokers were a good solution to authenticate users and guarantee quality. Now, as we're poised on the brink of an explosive growth of Electronic Commerce (*Ecommerce,* estimated to top a trillion dollars per year within a few years), equivalent mechanisms to grease and protect the wheels of this new form of distant and anonymous commerce are even more important.

HEAD-SCRATCHING ISSUES   Consider these scenarios:

- We are about to purchase a product over the net, but who assures us that the product is actually as described? Who verifies that the electronic money we send is "good"? Who verifies that the Emoney we send is actually received by the merchant we think we're sending it to? How does the merchant verify that we're the one sending the funds? Who assures the privacy of the transaction?

- We search the Web for some information, but how do we know it will be up to date, correct, and credible? In short, how can we depend on the information?

[1] And today we see the emergence of guarantors in electronic commerce. See www.forbes.com/forbes/99/0809/6403092a.htm

[2] Their counters were called *banchi,* from which the word "bank" was derived.

- We send out some information via the net, but will it be received in its entirety, unchanged, and only by the people we send it to? Will we be able to confirm that it has been delivered? How can we guarantee the accuracy of the information we send? How can we assure that the people receiving our information will only use it in ways that we allow?[3]

- We allow our doctor to perform real-time monitoring of some of our vital signs, like an electrocardiogram, through a device attached to our body, perhaps to look for an abnormality during normal activities, and this data is automatically sent to the doctor's office. It goes into our history and may form the basis for treatment (perhaps even treatment delivered by the monitoring device). But how can we be assured that the information transmitted by our sensors was accurately received? How can we be sure that *our* data, and not someone else's, is sent to our file? We wouldn't want to receive treatment based on somebody else's condition. How can we be assured that the person analyzing the data and prescribing the treatment is really our doctor, and not someone "hacking" into the system?

These examples share several fundamental issues that need to be addressed, or at least we need to be aware of them, to safely navigate the new ways of living and doing business in the Knowledge Age. We need to be able to assure the integrity and accuracy of information, know just where it came from, assure that it arrives to the people intended and in the form originally sent, and know that the transaction will be private and secure.

NEW SOLUTIONS, NEW PROBLEMS

New technologies are emerging that will address many of these technical issues,[4] but there are some issues that lie beyond the technology, which are part of the new business and communications context of the Knowledge Age. It's not just about new technologies that allow us to do, at a distance, what used to be done in person; it's about learning new rules for a new game.

Let's use Carnegie Mellon University's prototype "ask the expert" service[5] as an example and imagine trying to sell a commercial variation.

As the technology continues to improve, it will reach a point where it may be difficult for someone to be positive that they *are* talking to another human! But if they *believe* they are talking to a real person, there might be

---

3 The music industry is particularly embroiled in this issue at the moment, as they struggle with how to allow their intellectual property, the music, to be distributed electronically without massive illegal copying. Their response is the "Secure Digital Music Initiative," or SDMI. See www.sdmi.org/

4 See focus sections "Ownership and Security Technologies" and "Technologies for Electronic Commerce."

5 With this service, we can interview the computer simulation of a famous personality, seeing them on video while we speak to them, and then hearing their answers. See www.grandillusionstudios.com/webdemos.htm

some rather negative results if they act on possibly incorrect information that was generated by the well-meaning computer simulation. For example, suppose someone had a conversation with what they presumed was a high official in their government, and reacting to the discussion, the computer simulation suggested they "take the law into their own hands." Needless to say, this could be interpreted in several harmful ways, and where would the liability lay if this person did go out and commit an illegal act? Clearly it would lay with the perpetrator. But would it also rest with the programmer, the company that provided the simulation, and the simulation's content provider? Indeed, haven't we seen some of these concerns surface in relation to violent video games and movies?

The boundary between reality and fiction is very thin and wavy. It might be benign, for example, to interview a simulation of Einstein to learn about relativity, but it would be considerably more risky to discuss a medical condition with a seemingly real doctor, and then take action (or not) based on its recommendation. Or, suppose you had a virtual business meeting over the Internet and took action based on the decision of the person you were meeting with, only later to find out that it was her simulacrum—she was asleep on the other side of the globe at the time. Perhaps, as computer simulations become more believable, they will have to be marked in some distinctive way. Although, might they then one day form a lobby to press for "simulacrum rights"?

The Global Village Doesn't Support Local Ties

Of course, one obvious answer might be restrictive legislation regarding such simulacra, especially in the global environment of cyberspace; local or even national laws have questionable jurisdiction and could be "interesting" to enforce.

TRUST IS EVERYTHING   There's another issue, that of credibility, which goes beyond the authenticity issues we just explored. Today, we associate a level of credibility with various publications. For example, you might tend to believe a seemingly disruptive fact read in *Physics Review,* whereas you might be less inclined to believe the same information in the supermarket tabloid. Similarly, newspapers and TV news shows have developed their own levels of credibility, and that model has worked fairly well when the number of information outlets typically available to any person was rather small. Today though, through the Internet, there's a vast array of information on almost any subject, and search engines or pointers may well take you to sources you've never heard of before, which most certainly are not the "household names" of the news sources we've previously relied on. Anyone can post information to the Internet, and unlike in distant history (a few years ago), today, a business of one person can put on a Web face that rivals that of long-established, multinational corporations. So how to decide what to believe?

The Role of the Editor Won't Disappear but Instead Become
Even More Important

Although the Knowledge Age is accused (rightly in some cases) of causing disintermediation (removing the "middle man" by letting anyone get right to the source of things), "editors" are becoming far *more* important than ever before, because they can act to consolidate information with a given level of trust. An editor of a Web site, for example, will, over time, demonstrate his or her credibility in much the same way that dead-tree publications of the past have done.[6] Therefore, if you were looking for information about a particular subject, you might choose to stay one level away from the "great unwashed masses" of information, at least initially, and let a trusted editor's publication do the first level of "weeding out." In fact, in a world of free basic information, you might well choose to pay for such a service.

The Internet is also spawning completely new ways to help us determine the credibility of information. For example, most Internet Service Providers (ISPs) allow their subscribers to set up "discussion forums," where they can carry on conversations with like-minded individuals to discuss a particular topic.

Going even farther, a new company, ThirdVoice,[7] has harnessed the power of the Web in a way that is so simple, but powerful, that we may wonder why we didn't think of it first. The goal of ThirdVoice is to allow anyone on the Web, after downloading a small bit of software for their Web browser, to apply their own annotations to *any* page, regardless of who owns it, anywhere on the Web!

Now, at first, this would seem impossible, because few Web publishers (intentionally) allow the general public to change or deface their Web pages. But ThirdVoice maintains the annotations on its server and overlays them on the browser window as if they were a series of yellow

---

6 "All the information that's fit to print," *The New York Times.*

7 See www.thirdvoice.com

"sticky notes" on a sheet of transparent plastic on top of the original Web page. So whereas the original page is not, in fact, altered, if you've installed the ThirdVoice software, you will see everyone else's comments about the content on the page—instant global discussion on the credibility of the information! In fact, if you publish a Web page and you haven't installed ThirdVoice, you might be rather surprised by what may be a virulent discussion going on right on top of your work! Try *this* with an Industrial Age book.

Of course neither discussion groups nor ThirdVoice-style commentary is a panacea; a popular site or topic may generate so much discussion that just sorting through the comments, many of which may be irrelevant, may be an almost impossible task.[8] In fact, new classes of programs are emerging that may try to filter such discussions (a difficult and perhaps dangerous task even for human moderators).

Of course today's model of such Internet-based discussions are bound to change. For example, today most discussions are between people sitting at keyboards, but as "intelligent agents," or more complex computer simulations as we discussed above, become more common, you may be carrying on a discussion with a program, not a person. Perhaps even worse, the Web might become cluttered with a vast array of discussions between silicon entities, with no humans involved at all. Here's a scary thought: suppose we "mere humans" found value in the result. Indeed, suppose such a computer-to-computer discussion developed a new patentable idea—are the world's patent offices ready to decide who then could own the patent?

On the positive side, as our programs get better at understanding the content we (or they) place on the Web, we may also develop the ability to tailor that information to the viewer. For example, information on ancient Egypt could be presented one way to a scholar, in an entirely different way to a sixth grade student researching a report, and in yet another way to a tourist who just arrived in Cairo.

Of course keeping things current on the Web is almost impossible, especially when documents link to other sites, and as with any difficult task, that open up the opportunity to provide a valuable service that people will pay for (or that will differentiate a product). For example, the online Encyclopedia Britannica uses their guarantee to continually verify the accuracy of any pages they point to, even if not their own, as a selling point.[9]

8 According to the 1997 estimate of one company that acts as an archiving service for thousands of discussion groups, *Deja News* (**www.deja.com/**), about two-thirds of the messages that appear in discussion groups are irrelevant to the subject at hand.

9 Of course even the venerable encyclopedias of the past are changing. According to the July 30, 1999, Mike Elgan's Win Letter (**www.winmag. com/people/melgan/winletter/47. htm**), the *Encyclopedia Britannica* will no longer print a dead-tree tome—it will now only be available as insubstantial bits on a CD-ROM or over the Web and since October 1999 it is available on the Web for free at **britannica.com**

With all this information out there just waiting to be tapped, there's an obvious opportunity to profit from acting as people's guides and that's just what the Web search engines, and their successors, the "portal sites," have done. They derive their profit from advertising and sometimes "positioning" of how particular sites are ranked.

INTELLECTUAL PROPERTY There's another interesting issue surrounding what we might term "complex documents"—those Web-based discussions, annotated pages, and pages that link to others' pages—as to what constitutes the "complete work," and who owns it. It may not seem important today, but suppose, for example, someone used an Internet discussion group as the basis for a movie script, and this came to light. Would the individual participants[10] in the discussion group have ownership in the resulting screenplay? The Internet is sure to keep the lawyers busy for a very long time.

Hear Them All and Choose One

In this section, we have dealt primarily with issues of information authenticity and credibility, but in the Knowledge Age, almost everything is "information"! Even material goods, like household appliances, are presented on the Web in terms of information (both, as we've seen, by sellers and consumers), and the issues of authenticity and credibility apply here as well.

For example, some sites are manufacturers' sites, displaying their products in the best possible light. Then there are seemingly independent sites where consumers can educate themselves about products across manufacturers, search for those that interest them, and then perform a side-by-side comparison of the features they want most. CompareNET is one example, which one of your authors would have liked to have known about before recently buying a new stove.[11] Of course the importance of the accuracy of the information presented on such a site—and its credibility—is obvious.

On the other hand, there are also sites, such as ProductReview Net,[12] that focus on the subjective reviews of sometimes satisfied, sometimes disgruntled buyers, where you have to recognize the subjective nature of the comments. We now have unexpected combinations of the two, using technologies such as from ThirdVoice, which we discussed above, where users add their own comments "on top of" otherwise authoritative sites.

---

10 Note that there are technologies that can help maintain a "stream of ownership" for multimedia elements, called "watermarking." Basically, these techniques embed normally invisible information into an element that can be decoded to show its ownership. See www.cselt.it/ufv/products/watermark/talisman.html

11 See www.compare.com Another example is MySimon (see www.mysimon.com).

12 See www.productreviewnet.com/splash.html

Taken together, these sites represent a very broad and comprehensive "information set."

To date, there haven't been widespread centralized "certifying authorities" providing "certificates" that declare a site's authenticity or credibility; it has been more in the hands of the users to make those determinations. But we are now seeing some traditionally trusted authorities, such as the Better Business Bureau,[13] extending their purview into cyberspace.

The bottom line is that in the Knowledge Age, where we can all choose to be active participants instead of simply observers, the opportunities for business multiply. But, as when money was first introduced and counterfeiters flourished, every buyer should beware.

Now, let's move on and explore some of the (legal) ways that we might "choose to become rich" in the Knowledge Age.

[13] See www.bbb.org/library/cybershop.html Another site with advice is www.bizrate.com

Ownership and Security Technologies

Five-thousand-year-old relics found in the tombs of the pharaohs still bear the name of the person who made a specific vase, a painting, or who wrote a papyrus scroll; so it's easy to determine who made what. On the other hand, the first known cases of counterfeiting also date back to that same time period, so the water of authenticity is a bit more muddy than it might have appeared. Those difficulties are with *physical* objects—when we enter the realm of intellectual property (unsubstantial works, like music, poems, or formulae), it can be even more difficult to determine their paternity, because one work may well have been influenced by, or based on, prior work. For example, Isaac Newton and Gottfried Willhelm Leibniz spent several years bickering back and forth over who was the father and who was the plagiarist of infinitesimal calculus. Three centuries later, a close examination of the documents indicates that they were both right (each could claim paternity) and both wrong (neither copied the other).

INTELLECTUAL PROPERTY PROTECTION In the Knowledge Age, intellectual ownership assumes an ever-greater importance, because the proliferation and value of non-material goods is on the rise, and such goods can be most difficult to protect.

Although the patenting process for hardware has been well established, protecting software remains a fuzzy area that is evolving and being defined by many governments right now.

Protecting information itself, as well as the specialized services and applications that add value to the information and those that make it available and easy to access, required a constantly changing set of technologies. Additionally, each society maintains different views as to what constitutes acceptable behavior.[1]

Some of the techniques for protecting software show great ingenuity. For example, when distributing music files over the Internet, an obvious concern of the music owner is that someone will buy the song and then Email it to hundreds of his friends. One technique to combat this is to not send the music file in a commonly playable format, such as MP3,[2] but to embed

---

1 For example, it was reported in July 1999 that 98% of software used in Vietnam is pirated, a far higher ratio than in some other countries.

2 See www.mp3.com

the specially encoded music file along with a player that will play it. What's the advantage? This player also contains the credit card information used to purchase the song, and it can be displayed whenever the song is played—a strong deterrent to someone for passing the song on to a friend, much less to an Internet distribution site!

How about protecting physical media, such as tape? Work underway at MIT is embedding special nanoscale ferrous oxide particles into the tape, which encode an ownership trail. These particles can't be read or even detected without special equipment; so these special particles can't be duplicated except at prohibitive cost. The end result is the owner can easily determine if a tape is an original (and if so, which one) or if it is counterfeit.

At the logical, rather than the physical protection level, elements of the new MPEG-4 multimedia standard (as well as some stand-alone programs) have the ability to encode ownership information directly into the audio or video streams. But instead of being added as separate information that someone could choose to remove when duplicating the file, this information is encoded directly into the audio or video information itself in a way that can't normally be seen or heard—or removed without degrading the quality of a copy. If a full copy is made, this "hidden in plain sight" "digital watermark" information can be retrieved with a special program, clarifying its ownership. It is interesting to note that these forms of "protection" assume that the files may be duplicated but then allow for someone to verify its ownership.[3]

Other approaches attempt to limit the distribution of a piece of intellectual property. For example, some software programs make use of counters to control the number of installations allowed. In one example, based on software distributed by floppy disk (which, unlike a CD-ROM, can be modified), a specially copy-protected floppy disk moves one of a few "tokens" onto a hard drive during program installation, removing it from the floppy. When the floppy runs out of tokens, it won't perform anymore installations until an "uninstall" process removes the software from the hard disk and returns the token to the floppy. There are always ways to "beat" any protection system—it's a constant balancing act between protecting assets and making it difficult for legitimate users to make use of a product.

Another approach for protecting intellectual property is to encode it in a way that is only useful to designated recipients. Of course this practice is hardly new; encryption has been used for ages and has had some significant impact on recent history.[4]

---

3 This MPEG4 watermarking technology is designed with the assumption that multimedia materials will be sent over the Internet, or broadcast over television channels. Therefore, a computer monitoring the file server or broadcast channel can decode the watermarked information and verify that only acceptable forms of use are allowed. But it's interesting to note that the encoding is done in such a way that only the owner, or her representative, can determine the ownership, thereby preventing a third party from recognizing, and trying to mask, specific ownership information. See **www.cselt.it/mpeg/standards/mpeg-4/mpeg-4.htm#E10E14**

4 You may recall the German Enigma machine, which mechanically encoded messages in a way they thought couldn't be broken. When the allies broke the Enigma code, it gave them a decided advantage. This was not a one-way street—the Germans later broke an English code giving them a similar advantage.

Rather enhanced versions of these old encryption techniques are heavily used today in commerce, such as banking transactions, to prevent a third party from being able to use any data it may intercept, and to prevent a third party from inserting its own bogus transactions into an otherwise valid stream. Two techniques used are "secret code" and "public code" cryptography.

Secret, or symmetric cryptography, uses the identical code for both encrypting and later decrypting a message. In theory, if you don't have that code, you can't successfully decrypt it.

One problem with this technique, though, is that you must have an alternate secure channel to send the code to the receiving party. Using public code encryption, however, gets around this problem. With this technique, two codes, rather than one, are used. The two codes are linked in a complex mathematical relationship, such that if you encode some information using your secret code (which you never give to anybody), anyone who has your public code, which you do freely distribute, can decode your message, but they can't use your public code to encode another message as if it had come from you.[5]

Secret code cryptography uses the same code to encrypt and to unencrypt. The sender and the recipient of an encrypted message must therefore share the code. Public code cryptography, instead, utilizes two codes: one to encode and the other to unencode. The two codes are linked by a mathematical relationship so that data encrypted with one code can be unencrypted only and exclusively with the other code; for this reason the user has two codes: one known to many and one only known to that user.

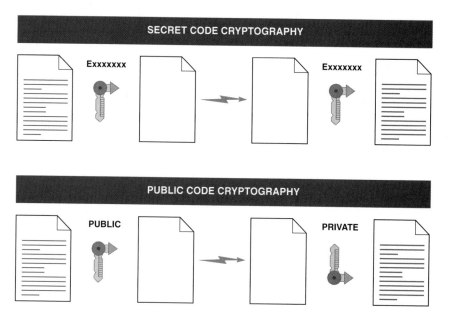

---

5 This type of encryption is based on the fact that it's very easy to multiply two numbers and get the results, but it's extraordinarily difficult, for large numbers, to take a result and figure out which two numbers had been multiplied to create it. (This process is called *factoring*.) In this context, "large" means a number with more than about 448 bits in the key. See news.cnet.com/news/0-1005-200-330045.html However, note that as processing power increases, it will become possible to break even these complex codes. Indeed, in January 1999, a special-built computer by the Electronic Frontier Foundation called "DES Cracker" broke the 56-bit DES code (the largest key then freely exportable from the United States) in a mere 22 hours and 15 minutes (see www.eff.org/pub/Privacy/Crypto_misc/DESCracker/HTML/19990119_deschallenge3. html). You may want to think about this before trusting sensitive data to such small keys (128-bit encryption, used in browsers within the United States and in some global instances, is, for the moment, still relatively secure).

TOO MANY NUMBERS TO REMEMBER These mechanisms, although they can be used to secure computer-to-computer transactions, don't lend themselves to day-to-day transactions such as logging into a computer, because most people won't remember long and obscure numbers. In fact, given the number of passwords that most of us have to deal with these days, we need some better mechanism to provide secure access to our information assets; one that takes into account what humans will find easy to use (so, for example, we aren't tempted to use the same password for every account, but we don't have to remember dozens of different passwords).

One contender is the "smart card"; it looks just like a credit card but it has a processor inside rather than a magnetic stripe. This allows the smart card to be programmed to store many different passwords and to communicate them securely to any computer that has a smart card reader. It could also, for example, replace your car's ignition key (and the key of your rental car—the rental company would just download a temporary "key" into your smart card at the time of the rental, which would expire when the rental period was over). And you can download "cash" value into the card from your bank account through an ATM or an appropriately equipped PC—or even through a cellular phone that has a smart card slot!

Smart cards are having trouble getting started in the United States. Unlike in Europe, where smart cards are in general use already, most people in the United States already have magnetic-stripped credit cards, which has slowed the move to this new system. However, significant growth is expected.

The problem with the smart card, of course, is that it can be lost (or stolen), so it's still desirable to add "authentication"—some method to guarantee that "you are you," especially when using your smart card.

A number of "biometric" techniques are currently in limited use, including fingerprint recognition[6]; facial recognition, using a camera to compare the pattern of your eye's iris to "your" known pattern (which is even more unique than a fingerprint); and voice recognition. But it's still unclear which of these will become widely adopted.

We began this focus section by noting that it's difficult to protect unsubstantial intellectual property on the public Internet, but some things are making this more feasible beyond the technologies we've already discussed. For example, on the societal side, laws are (slowly) being updated from their Industrial Age roots,[7] and even more intriguing technologies are being explored, such as developing computer programs that can, in an automated manner,

---

6 For example, see **www.whovision.com/technology.html**

7 In 1997 an agreement was signed in the United States against intellectual property theft, the "No Electronic Theft Act" (**news.cnet.com/news/0-1005-200-325007.html**).

examine many different programs and discern whether they are similar enough to be "rip-offs."[8] (Interestingly, software such as this is also beginning to look at term papers and other educational output, trying to determine if they might be copies or modifications of the many such papers already accessible over the Internet. Who knows—some schools may begin to use such programs to watch for plagiarism within their schools, as well!)

---

8 Professor Alex Aiken of Berkeley has announced a program able to examine other programs and show their similarities. The program is called MOSS, Measure of Software Similarity, and is available via Internet: **www.cs.berkeley.edu/~aiken/moss.html**

8   A Vision of Tomorrow: "A Million e-Businesses, Interacting with a Billion People, through a Trillion Interconnected Intelligent Devices . . ."[1]

Is "Electronic Commerce," or "Ecommerce," real, or is it a product of the Internet's hype?

There are thousands of individual statistics out there from hundreds of groups; some building on each other to huge end results and others scoffing at any meaningful changes that the Internet may bring to how people, and businesses do business. So let's begin this chapter with findings from one well-respected publication that doesn't have any direct ties to the success, or failure, of the Internet—The Economist[2];

A recent worldwide survey of 500 large companies carried out jointly by the Economist Intelligence Unit (a sister company of The Economist) and Booz Allen and Hamilton, a consultancy, found that more than 90% of top managers believe the Internet will transform or have a big impact on the global marketplace by 2001.

[Forrester Research] argues that e-business in America is about to reach a threshold from which it will accelerate into 'hyper-growth.' Inter-company trade of goods over the Internet, it forecasts, will double every year over the next five years, surging from $43 billion last year [1998] to $1.3 trillion in 2003. If the value of services exchanged or booked online were included as well, the figures would be more staggering still. . . .

[Then,] Britain and Germany [will] go into the same hyper-growth stage of e-business about two years after America, with Japan, France and Italy a further two years behind.

That's a heady prophecy, and if it comes to pass, it clearly offers tremendous riches to those who stretch their established business practices

---

1 Paraphrase of comments by IBM CEO Lou Gerstner, as presented by Al Zoller at The Economist's "New Directions In Computing" conference, London.

2 "*The Net Imperative,*" The Economist, June 26, 1999. www. economist.com/editorial/freeforall/ library/featured_surveys/index2_ 19990626_su9828.html

into this new way of doing business. But it also portends a "changing of the guard"—established companies who feel they can keep a lid on Pandora's box are likely to crumble under the pressure.

But what's the allure? Why are people rushing headlong into a new way of shopping? Why are businesses tossing out traditional purchasing processes to buy and sell between each other online? (Note that "business-to-business" transactions in 1999 represent far more dollars of Ecommerce than "people-to-business" electronic retail, or "Etail" Ecommerce.) It's all about that triumvirate: time, convenience, and money.

ARE THEY BUYING ON LINE? Currently, Ecommerce is most at home in the United States, with Europe a distant second. But why? It's not a technology issue; the same chips, fibers, and applications used in the United States are also available in many other countries. Instead, it's a matter of culture,[3] and cultural changes take far longer to implement than mere technological ones. For example, it's not that unusual for a European to do business with a shop that their parents and grandparents did business with, and that personal relationship, normally missing in the United States, is a strong incentive to not shop through the Web.

Additionally, Europeans have historically not embraced catalog shopping to the extent of Americans. Because what we've seen on the early Web is, in many ways, a conversion of traditional catalog and shop-by-phone shopping into a more convenient, and sometimes a more interactive form of electronic shopping, those countries with people who enjoyed letting their fingers do the shopping were more willing to let them walk from the telephone keypad over to the keyboard. Nevertheless, as the Economist suggests, it seems likely that "progress" is coming to Europe.

Europe is likely to be catching up in the Ecommerce arena in just a few years. Why? One reason is that a growing number of European companies are "going global," finding value in selling merchandise to a worldwide market while also purchasing their raw materials, plus consumables like office supplies, from the global marketplace. In this global marketplace, the ability to tie to their customers, and suppliers, electronically—instantly—becomes a competitive necessity. (For example, in the United States in 1998, business-to-business Ecommerce was ten-times greater than business-to-consumer Ecommerce, because of the competitive value this conferred to businesses.) So as more European businesses embark on

3 More than 100 boutiques in nine European countries took a challenge from Microsoft, Hewlett-Packard, Visa, MasterCard, UPS, and other companies during the 1997 holiday shopping season, to set up the first European "virtual shopping center." It was called "E-Christmas (www. e-christmas.com). The shops provided attractive content, and the technology companies provided a sound platform for Ecommerce and for the delivery of the purchased goods. The result, though, was less than satisfactory; out of 184,000 hits on the Web site, only 360 sales were made! The folks in Europe, at least then, used the virtual shopping center to get ideas, and then they went to their local shops to do the actual purchasing.

business-to-business Ecommerce, and as people recognize the advantages that business-to-consumer Ecommerce can have, it seems likely that European Ecommerce will continue to pick up speed.

ECOMMERCE IS CHANGING THE RULES

Consider this: the distribution chain for traditional retail products is often long and complex, going from producer, to sometimes several levels of wholesalers, and finally to retail shops. This chain gives customers very little direct interaction with the actual manufacturer, which isn't good for the consumer or for the producer. With Ecommerce though, the manufacturer can get direct feedback from its end customers (sometimes more than they might want if they disappoint them). But for businesses that really do want to satisfy their customers, this direct feedback (if they pay attention to it) is invaluable. From a consumer's standpoint, they benefit from such interaction, and they also suddenly find themselves in an unusual position of power—using the Web, consumers from around the globe can now band together to negotiate lower prices, something traditionally reserved for wholesalers and large retailers![4]

NAME YOUR PLEASURE, THEY CAN SELL

Of course the Web abounds with information that you can slice and dice to your heart's content. Want a list of all TVs under $500? Many sites will be happy to oblige. Want a vendor-neutral comparison of different vendors' wares? Many sites, such as Compare.com,[5] put such information a click away.

Or, suppose that consumers were able to set the value they are willing to pay for a product—not in the aggregate of the marketplace, but individually. On the Web, it's not impossible to turn the traditional retail model on its ear, as demonstrated by Priceline.com.[6] There, you can say, "I'll pay so many dollars for an airline ticket from Boston to Milan." Then Priceline will offer your bid to airline companies and ticket resellers to see if anyone's willing to accept even a ridiculously low price. Which they may well do if the seat would otherwise go unused—there are few goods more perishable than an empty airline seat when the cabin door closes!

Indeed, the new business models that we'll see emerge are limited only by our imagination. For example, would you like to be paid for surfing the Web? CyberGold will be glad to oblige.[7] Or, would you like a coupon to cut the cost of something you're planning to buy? CoolSavings will let you print perfectly legal coupons right from your PC.[8]

4 Mercata (www.mercata.com) is a company started by Microsoft founder Paul Allen, to do just this—it allows consumers to use their aggregate buying power to tell merchants just how much they're willing to pay for a specific product—and the merchants, watching the number of interested buyers climb, can lower their price until the "twain do meet." Truly "power to the people."

5 See www.compare.com/

6 See www.priceline.com

7 See www.cybergold.com/

8 See www.coolsavings.com

FROM ME TO YOU   Interestingly, the Internet is not just about retail transactions between consumers and businesses, but also about transactions between consumers. Rather suddenly, "auction sites" such as eBay[9] have sprung up to act as intermediaries between consumers wishing to buy and sell virtually anything under the sun. Which is interesting, since many people felt the Internet would destroy intermediaries—instead, while some "Industrial Age" intermediaries who no longer add much value are indeed losing out, other, "Knowledge Age" versions, such as the auction sites, are generating significant new wealth.

Flea Market—Garage Sale

Indeed, even workers and jobs are being "sold" on the Web; an increasing number of employers expect applications to be filed on the Web, and most college graduates now put up their own Web page, offering that address to perspective employers as an extended resumé.

Online job hunting can also go beyond typical job search sites. Recently, a complete team put itself out for bid on the Web! Perhaps these new directions are a good thing, because the traditional "cradle to grave" work contract with large companies is increasingly a thing of the past—it's said that most employees joining the workforce today will have three or more jobs during their work life.

Another change we're seeing is the shift from products to services,[10] with the "free PC" movement being a good example. As with the cell phone model before it, a growing number of manufacturers and Internet Service Providers are offering to give customers a reasonably configured PC for free—if they agree to sign up for two to three years of Internet access. Of course the PCs aren't really free, but some people perceive them that way, and are willing to pay a small monthly fee rather than pay for a PC and still have to purchase Internet access.

There's another similar emerging trend—that of the Application Service Provider (ASP). Rather than going out and purchasing an expensive software program (such as an office suite), new companies, wishing to gain more of a customer's business, are offering to "rent" them the use of the software, which resides on their (the ASP's) servers, as needed. Although there are still issues to this, such as the need for fast connectivity and some questions about the security and resiliency of off-site data, it

---

9 See www.ebay.com

10 According to IDC the selling of services over the Internet will grow from the $7.8 billion in 1998 to $78.5 billion in 2003.

demonstrates but another way that people are experimenting with new, Knowledge Age business concepts, as every business struggles to demonstrate its competitive advantage.[11]

Of course there is definitely such a thing as information, or shopping, overload. With so much competitive information, and so many places to turn to shop (and if you think there are a lot now, just wait!), it's becoming increasingly difficult to figure out where to go: enter—the "Portal."

Shop until You Drop

Typically based around one of the established search engines or other high-value sites, a portal attempts to offer you so much customized information, with a given level of "quality," that (they hope) you'll make it the "start page" for your browser or, at least, that you'll visit often. Why? Based on the number of visits ("hits"), they get to sell advertising and "real estate" to those firms that would like you to see them first.[12] If done right, it's like a mall—a "known quantity and quality" of vendors and information.

FURTHER INTO THE FUTURE Of course ideas on bringing information to people are likely to change as extensions to today's technologies, and new ones, emerge.

For example, it seems that Professor Vincentelli at Berkeley University is working toward tiny dust-speck-sized devices that actually use helicopter-like rotors to remain airborn even without wind.[13] These autonomous devices, called "Smart Dust," will contain a thick film battery and solar cell, a DSP (Digital Signal Processor), sensors, and the ability to communicate with other active things around them—other dust motes, and with the next generation of cell phones, notebooks, and other appliances that may sport short range radio (such as Bluetooth).[14] Imagine what they might communicate! (But then, there might be all that additional dust to be cleaned up—could that generate a spin-off business of very high capacity filters to suck these devices out of the air? Can you say *Technological Escalatio*?)

We began this chapter with the quote, "A million e-Businesses, interacting with a billion people, through a trillion interconnected intelligent devices," and as we can now see, there really may be that "trillion interconnected devices." Now, just imagine the changes they may bring.

11 See "Suddenly, Software Isn't a Product, It's a Service," June 21, 1999 *Businessweek,* www.businessweek.com/1999/99_25/b3634016.htm

12 But this may not be a panacea. On one hand Forrester Research estimates that in 2003, the top nine portals may get 20% of Internet traffic, which is growth compared to the 15% they received in 1998. However, the number of people clicking on ad banners has decreased from 2% in 1997 to only 0.5% in 1998.

13 See robotics.eecs.berkeley.edu/~pister/projects.html

14 See www.bluetooth.com/v2/default.asp

Your car may notice something going wrong and query each service station it passes until it finds one capable of fixing the problem; perhaps even putting you, the consumer, in charge by choosing the service station that will agree to charge you the least for the service! As we've seen, this idea of our "agents" doing such shopping is not science fiction.

But there's another synergy that might evolve out of the increasingly dense soup of information, people, and their agents—might it soon become easy for each of us to create our *own* services, for our own use as well as to sell to others? Could we, the consumers, also turn into vendors? This, and more, we'll explore in the next chapter.

Electronic Commerce

Tired of fighting traffic to reach the mall? Of standing in line at the cashier? Of walking past endless shelves looking for something that might be hidden somewhere? How about being able to buy something without having to pay a sales tax? Ecommerce is a popular answer, being embraced by 25% of Americans as of 1998 and expected to grow to 50% of Americans by 2000.[1] Both individuals and companies are moving to an electronic marketplace.

On one hand, Ecommerce can be seen as a logical extension of catalog shopping over the phone, although the Web opens up a global catalog. However, Ecommerce is more than a catalog. It is something different that requires a new set of infrastructures and related technologies. Let's take a look.

In an electronic marketplace, issues such as validating both the buyers and sellers and assuring that transactions are sent faithfully take on new importance. Additionally, the secu-

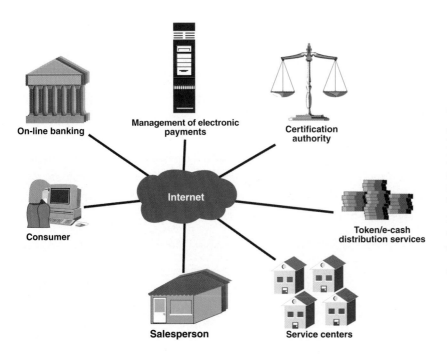

On-line banking

Management of electronic payments

Certification authority

Internet

Consumer

Token/e-cash distribution services

Salesperson

Service centers

The different roles of electronic commerce. Of these, only token Ecash distribution services, on-line banking, and the management of electronic payments are specific to electronic commerce. The others are also present in normal commercial transactions, even though they are not often evident.

---

1 Although the tax-free status of the Internet is likely to change over time.

rity of the information about a transaction, and about the consumers themselves (especially across multiple electronic purchases from many vendors) becomes important to safeguard.[2]

This is an important area to consider—look at the difference between the United States' most common cellular phone format, which is easy for anyone to listen to using a scanner and Europe's GSM technology, which encodes the voice in a way that's much more difficult to intercept.[3]

THE ELECTRONIC CASH   The way we pay for things electronically can also be an issue. For example, today, most Web-based transactions are paid for with a credit card, but that clearly identifies who bought what. In the physical world, someone wishing anonymity can pay with cash. This is a "freedom" that some find missing and important in the emerging Ecommerce world, but there are currently no widely accepted forms of "electronic cash." Note that in the "real world," there are "third parties" who guarantee the monies being paid—a credit card company once they've authorized the transaction, or a central government, who stands behind its cash—but these are institutions that have yet to be fully integrated into the electronic marketplace.

"Common wisdom" is that people hesitating to shop on the Web do so because they're afraid their credit card information may be stolen and used fraudulently. Now this can, of course, occur, but why would this be more likely than if you hand your credit card to an unknown gas station attendant who disappears inside to make your transaction? In fact, your information, encoded in electronic form, may be much more secure on a vendor's server.[4] Interestingly, credit card transactions may be more secure than using cash—common ink jet printers are now so good that in 1997, 43% of counterfeit U.S. bills were printed on PCs![5]

Another advantage of Ecommerce is the extremely low cost per transaction. One experimental form of Ecommerce payments developed by Digital Equipment Corporation is Millicent, which lowered the per-transaction cost to 1/1,000 of a penny,[6] enabling transactions that could never justify the traditional per-transaction fee charged by credit card companies. If such low-cost payment systems catch on, they will enable a vast array of "one-off" services, such as paying a few cents for the translation of a few sentences from English to Japanese. This could be far less expensive than buying a dictionary!

---

2 Indeed, many countries are beginning to address these important issues, such as the United Kingdom's work in 1999 on the Electronic Communications Bill—www.wired.com/news/news/email/explode-infobeat/politics/story/20937.html

3 Of course there's a "dark side" to this force—it also implies that criminals' conversations are more difficult for law enforcement agencies to intercept.

4 Of course as Ecommerce is maturing, there have been some notable security lapses, with un-encoded credit card information being unintentionally made accessible on the Web. But even in such cases, your liability is probably limited in the same way as when a credit card is stolen. Additionally, a growing number of credit card companies are holding its cardholders completely harmless from Web-based fraud.

5 See news.cnet.com/news/0-1005-200-327989.html

6 See www.millicent.digital.com

Electronic payments are not limited to the Web. In Europe, for example, a growing number of countries are supporting the use of "smart cards" for paying for things like newspapers and subway and bus fares. In some countries, such as Italy, you can use your smart card to gamble over your pocket cell phone!

ELECTRONIC TRADERS Of course it isn't only people buying and selling electronically. Some industries, such as the stock market, are seeing a growing number of "electronic traders," which are programs that watch the ebb and flow of the market and trade according to often complex rules. One problem that has developed is that under the wrong set of circumstances, these programs can reinforce each other and actually cause market variations that have no other reason for occurring. Today, these businesses do maintain electronic and physical "panic buttons" to halt such runaway trading, but as the number of electronic traders continues to increase, it may be increasingly difficult to recognize when they're at fault and take action.

What about responsibility? If you make a purchase, in person or electronically, you are usually held responsible for your actions. But suppose your autonomous Intelligent Agent makes an error and purchases something at a price that you, later, consider far too high? Are you still responsible? Note that these are simply old problems in a new context—if your car fails and causes an accident, you are probably still held responsible, but we're "used" to such situations. We'll soon have to get used to their Knowledge Age equivalents.

THE ELECTRONIC SHELF Another important issue in Ecommerce is "visibility"—how to display products or services so people can explore them in a way that will lead them to want to buy. Many sites today use a catalog approach, displaying goods in "departments" or using other groupings that make sense. A shopper can add items to their virtual "shopping cart," typically with just a click so as not to interrupt their shopping experience, and then "check out," paying for all the items when they run out of steam (or money). But this interface—how easy it makes it for the customer to find merchandise and complete the purchase process—can easily make or break an Ecommerce site. Unlike in the "real world," where there is a penalty in walking or driving time to go to the competition, on the Web a competitor, even one across the globe, is but a click and a few seconds away. If you frustrate your customer, he or she is very likely to go elsewhere.

FOCUSING ON THE CUSTOMER There is, of course, no "right" answer to the best interface, and so merchants are experimenting with various types. For example, one school focuses on "personalization," keeping track of a shopper's profile of interests (either by having the shopper enter them manually or by learning from their browsing and shopping) and automatically offering up goods and services that may be of specific interest to this shopper. Future electronic agents, acting on our behalf, may be able to easily communicate this information to our merchants of choice. We may be headed towards a world of very targeted advertising.

On the other hand, there's the "interactive" model of shopping, where sites attempt to provide not just information, but an "experience" that goes beyond a mere electronic copy of a paper catalog. For instance, a site might allow us to download a model of our living room onto the Web and merge that with a new couch we're considering buying—a business that provides the most effective visualization tools might find they have a competitive advantage in the sale. Other sites may extend the old sales models, allowing you to use online tools to customize a product to meet your needs, such as changing the color of a piece of china or the colors in a fabric. (If you have created something new, might you then be able to offer it back to the store to sell to others, taking a cut of the profits?) Additionally, some sites are looking to merge the best of the new and the old. Say you've decided you like this new couch, but still have a few questions about the fabric that aren't answered on the interactive Web site. You may find a button that says "Call Me," and if you push it, a salesperson from the store would immediately speak to you either over the phone, or more likely by using "IP telephony" directly over the Internet. Just as with good telephone call centers, your profile might be right in front of the salesperson to help guide your interaction.

9   Do-It-Yourself Services

In the last 50 years we've become accustomed to buying ready-made products, trading off personalized goods for a lower price, and constant availability. In fact, the Industrial Age, with its techniques of mass production, has provided an enormous increase in the variety and quantity of goods available while bringing down price—but at a loss of some individual choice. For example, historically, shoes were made to order, directly fit to each buyer's feet. Of course only the wealthy could afford shoes. Today, most of us do have shoes, but we buy them for a (relatively) low cost "off the rack," choosing from the store's selection to get the best look and fit. It may not be a perfect fit, but most of us don't choose the alternative—custom, hand-made shoes, which can easily cost $1,000 and often take a month to be made.

However, the Knowledge Age is beginning to augment mass production with "mass customization"—using automation plus, in some cases, the ability to whisk information around at near the speed of light, to give people more precisely what they want but at a cost and in a time frame they're willing to accept.

JUST FOR ME   For example, consider Digitoe,[1] a company that delivers just such custom footwear for your tired feet. You first go into their store or a regional "scanning center,"[2] where a computer scans each foot to produce, in effect, a 3D model. The data are transmitted to Digitoe where the solid form around which the shoe is made, the "*last*," is created to your exact measurements.[3] At the same time, the leather for the shoe style you ordered is cut, based on the size of your lasts, and your shoes are made. The first time, with the creation of the last, your shoes can take 1–2 weeks. Subsequent orders, which you could conceivably make by looking at new shoe styles on the Web, can take a few days or less. In fact, you can design your own shoes (a market of "one"), with the assurance that they'll fit!

[1] www.digitoe.com/

[2] As of August 1999, available in Seattle and Port Townsend, Washington.

[3] And if your feet are sized differently from each other, this removes the problem of one shoe fitting and the other hurting—each shoe for each foot is custom made!

Of course it *will* cost more,

"It is more expensive to make shoes one pair at a time. [But] with the aid of modern technology and the CAD/CAM Computer, the cost difference is not very much."

However, the cost differential between an "off the rack" shoe and a mass-customized shoe should not be nearly as great as for custom footwear created the old way . . .[4]

There are many other examples of how new Knowledge Age capabilities will allow you to customize products to meet your individual needs, such as configuring your own car in the comfort of your den instead of in a high-pressure car dealership; or building your own pizza to be delivered shortly from a local pizzeria—you don't necessarily know or care from which one as long as it gets there hot and on time.[5]

Personalization: Another Opportunity for Transforming Ourselves from Spectators into Actors

"Personalization" is a key component of a growing number of consumer products and services. For example, some pocket cell phones allow you to download whatever "ring" sound you wish, and some cars will auto-position the mirrors, driver's seat, and steering wheel as you last left them, regardless of who drove the car in between (it identifies the main drivers by their specifically coded keys). Looking at a personalized service, consider your phone number—some services allow you to publish a single number that may have no relation to where you work or live, but that number will "follow you" to your cell phone, home, office, or wherever you are. Through the magic of DSPs (digital signal processors—very powerful, specialized chips) your stereo can personalize your den, transforming it acoustically into anything from an intimate restaurant to a concert hall.

Getting customized news is no longer news; many "portal sites"[6] and newspaper sites allow you to specify the types of news and weather information you are most interested in, and will automatically prepare your own "front page" each time you log in.[7]

Mr. Smith's "Daily News"

As we move toward the future, this type of information may follow us into our Internet-connected cars; we'll see this as our GPS navigation system being automatically configured to bring us to our next scheduled appoint-

---

4 Looking forward, if the "scan" of your feet were stored in a standard format, you could then send the model of your feet to multiple manufacturers and see what each could do for you. And why limit this to feet? How about a chain of "full-body scanning booths" showing up at the mail, where your 3D measurement is stored on the Web (under your virtual lock and key) to be used by clothing manufacturers of your choice to show you how their garments will look, not on a mannequin, but on YOU! Sound like science fiction? Its not—Adrian Hilton has just completed the first such full-body scanning booth at the University of Surrey! Although this booth takes some shortcuts because its intent is to generate "avatars" (computerized representations of you) for use in online environments, it is a very real first step. See **www.eurekalert.org/ releases/ns-sia072099.html**

5 **www.pizzahut.com**—but currently only in designated test areas.

6 Common starting places on the Web such as **www.yahoo.com** and **www.altavista.com**

7 See **my.yahoo.com/?myHome** for one example.

ment (found from our online calendar), skirting the latest traffic jams that occur while we're on the way. On the way home, we might get reviews of the latest movies or TV shows we are interested in seeing that evening, or perhaps we'll receive interesting recipes that can be made with the food that happens to currently be in our refrigerator and pantry!

From a computer standpoint, personalization has been a growing way of life. Many programs have had "preferences" that allow us to tailor their look, feel, and operation in ways we prefer. This has been taken to a new extreme with Microsoft's Office 2000; its menus continuously tailor themselves based on how we work so we don't (usually) have to configure them ourselves. That's interesting, because there's a limit to the number of things we want to personally learn how to configure; we balk at having to learn to interact with too many programs that are too different from each other.

You may or may not be a fan of the current "windowed" computer interfaces, but although far from perfect, they have brought a significant amount of consistency to modern programs. Just compare them to the early DOS programs where, even if they simulated a graphic interface (or worse, when they did not), each presented quite a different face. Today, most savvy computer users can "intuit" their way around a new program with little help, and this is very important from the standpoint of businesses offering new services—people are much more likely to adopt them if they "know" how to use them without having to be trained or without a long learning process.

On the other hand, human–machine interfaces have to be adaptable. Have you ever used one of the common "wizards" that step you through otherwise complex steps? These are often very helpful the first few times, after which you may be thinking that "idiot-proof interfaces are only for idiots." Subsequently, you'd probably like a more streamlined way of performing the same action, so various levels of interactivity are offered.

There Are Two Types of People: Those Who Admit They Don't Quite Know How to Use All of Their VCR's Features and Those Who Lie. Both Need a VCR That's Easier to Use

But perhaps what we really need is an interface that adapts to our (constantly changing) level of understanding of any product. Perhaps we need our own "intelligent agent" that monitors our use of each product, adjusting the interface according to our needs. If this were to really work, it would have significant value to business by reducing the amount of training time we might need on a given application. In fact if done well, no formal training would be required at all, as our agent would give us just the assistance we needed at any moment.

FLEXIBILITY FOR THE BUSINESS   At a less personal level, we are seeing the emergence of services that tailor themselves to a user's needs, for example, the phone network. Some telephone companies are implementing "intelligent networks" that allow a business to reconfigure its telephone infrastructure as needed. For example, calls to a mail order company from a particular geographic area might automatically be routed to a call-handling center in the caller's area to capitalize on local knowledge. Also, while on hold, local ads might be placed in the "on-hold" music. Speaking of hold times, based on Caller-ID, the business may recognize a caller's business relationship, moving good customers higher in the queue, and of course once the call is answered, the complete business history of this customer could be displayed to the operator, allowing him or her to provide what seems to be very personalized service.

So far, such interactions between the many programs that might provide such services have had to be individually customized, making it inherently difficult to "plug together" different functions from different vendors. In the future, a combination of the (relatively) platform-independent Java computer language and the emerging Extensible Markup Language (XML) may make it easier for businesses to "roll their own" functionality.

The Services Network:[8] Productive Infrastructure of the Information Society

Of course there are more than a few problems in trying to cobble together multiple services from multiple vendors as one new service from you. For example, if you're an Italian-speaking expert on classical Latin literature, people in the United States may not have heard of you to benefit from your expertise. Even with the growing number of directories on the Web, if an English-speaking potential customer did hear about you, the language barrier might make you both hesitate to try to work together. Suppose, though, that as the Italian expert looking to sell your services into the English-speaking world, you were able to "hire" a service that automatically trans-

---

8 See news.cnet.com/news/
   0-1007-200-325850.html

lated any incoming English Emails into Italian, and your responses back into English. If this translation were done well, your market could expand! However, today, there are no reliable ways to tie such translation services, from different vendors, into your personal Ecommerce solution—even if you programmed something that did work right now, if the translation software vendor changed something, your solution might fail without warning. This is a significant challenge for the future of Ecommerce.

Once this challenge is successfully met, even greater opportunities open up. For example, if you've satisfied your English-speaking customer with your (translated) Latin literature recommendations, she may now want to hire you to act as her virtual guide through online Italian museums (which might well require the use of additional online services to provide the guided virtual reality experience). Another revenue stream. And if this were successful, she might then be interested in actually traveling to Italy to receive a guided tour of the real thing. Again, if you could reach out and incorporate other companies' services as your own (paying them a fee, of course), you might be able to act as her travel agent to book her trip. Yet another revenue stream.

These simple examples are, of course, just to stimulate your thinking about the rich streams of commerce that might develop if it were easy to combine services on a moment's notice. With these tools, it will no longer require the resources of a large business to bring new ideas to market—a business of one may suddenly be able to compete on the global stage.

Technologies for Usability

SO MANY FEATURES, TOO LITTLE TIME! Pick a computer program you use every day. Do you use all of its features? Do you even know all of the things it's capable of doing?

Many people use only a fraction of a program's features; some say as little as 10%, and that many of today's programs are "bloated." But "bloatware" isn't limited to software; have you used all of the cycles and dispensers in your dishwasher or washing machine? How about your VCR? Do you know all the secrets of your cell phone? If not, why not? Is it that we don't want all these features? Or, is it that although they might be useful, trying to remember how to use all these (often cryptic) features, or even remembering that they exist at all, is more trouble than its worth? The key word here is "usability" or the lack thereof.

CAN WE REALLY USE IT? "Usability" is an umbrella term for an array of characteristics that make a product or service easy to use and "approachable." For example, a book printed with a font too small to read without a magnifying glass would hardly be usable. But how many times have you struggled to read labels on some cell phones? There are no "magic bullets" for usability, but some informal guidelines, learned from watch-

The French designer Jacques Carelman in his series "Catalog of Unfound Objects," furnishes examples of everyday objects that are deliberately impossible, absurd, or misshapen.

Coffee carafe for masochists.

ing products succeed and fail, have taken root. They take into account the product (or service), the type of person using it, and the context in which it will be used. For example, it's usually a good idea to present a new product in terms of what someone is already accustomed to using; this bootstraps a person's appreciation for what the product can do, and leverages off of ingrained habits.

One way to determine what would be usable for a given product is to observe a large group of people interacting with it (but be careful—designing to the "average person" can be problematic as well, as there is no mythical "average person").

Another way is to put the designers into their potential customers' shoes—in this case, into baby shoes: UNICEF created an exhibit that recreated a typical home—but from a small child's point of view. Everything was scaled up so that an adult suddenly found that the corners of tables were strategically positioned to hit them in the head! (Unlike in the real world, these corners were made of foam so as to only bruise egos, not foreheads.) Looking at the walls, it was now obvious why those large holes had such an attraction for things (the holes, of course, were the now-large electrical outlets). The point, which was well-driven home, is that designing for one audience (adults) didn't take into account the very different needs of a different audience (kids) with their different perspective.

This also applies to many products. In fact, it gets more complicated, because an interface that's easy enough to be used the first time may become very cumbersome with familiarity. For example, do you remember how you needed a little extra help the first time you used your word processor or your banking program? Without some assistance, you might have been lost, or at least never found many of their best features, but later, such handholding was unnecessary.

Computing has actually taught us a lot about human–machine interfaces. Apple, building on research initially done at Xerox's famed Palo Alto Research Center, used graphic representations of everyday objects, such as a desktop, to provide a more intuitive way for people to interact with their computers. Although it's far from perfect, it's so much better for the average person than the cryptic command line interfaces of the past that Windows, and most other platforms, have now adopted variations.

USE IT, THE WAY YOU USE IT! "Context" is another important element of the interaction between people and things. For example, books were originally scrolls wrapped around sticks (their "binding"), and were read standing up with the unrolled scroll in both hands. Later, books were unrolled on large lecterns; sometimes each (rare) book had its own lectern. Of course with the mass production of books, things had to change; books become smaller, easier to hold, and easier to store in cabinets. As the technology (moveable type, paperback books, etc.) changed, so did the context in which they could be found (on subways, in airplanes, and in neighborhood libraries.) In today's context, books in the old scroll form wouldn't be too "usable"—imagine trying to read a scroll in a coach airline seat!

In a similar vein, although it's technically trivial to store the text of a popular novel on a notebook computer (and some are available in this format), most people find that this isn't a very comfortable way to read a book. Notebooks are designed for typing and interactivity, and usually aren't as convenient as a regular book. (Although one of your authors was once in a hotel room, reading a novel before going to sleep, when the hotel lost electricity; finishing the paperback book was out of the question, but he did satisfy his before-bed reading habit with some documents on his notebook computer.)

The two worlds of books and notebook computers are, however, destined to merge. Some companies, such as NuvoMedia[1] and Everybook,[2] already have electronic books on the market based on notebook technology. But MIT, Xerox, and other research organizations and companies are working toward "electronic books" that more closely mimic the "look and feel" of a traditional book. Conceptually, these might appear as a book-sized device containing some electronics in the spine, and pages of "electronic paper," or "Epaper."[3] This special paper may feel like regular paper, but its surface could be changed by the books' electronics. Text of a novel could be downloaded from the Internet or installed through a memory chip, and each page would then take on the text and pictures of that book. Want another book? Just download it and the pages change. Indeed, given the impressive increases in storage we're seeing (IBM recently announced a 340 megabyte disk drive that can fit inside a golf ball,[4] as well as a 9 gigabyte notebook disk drive), one Ebook might hold thousands of novels. Add to that a mechanism for pay as you go, and you might have the past few years of *New York Times* best sellers sitting on a small portion of your Ebook's storage, just waiting for you to decide to read them (you would pay for the privilege only when you used it). Of course such technology should be just as appropriate for displaying moving images—want to watch Titanic while sitting on a deck chair during your next cruise?

As we can see, these new devices are building on well-known, traditional devices (with some twists, of course,) so people will innately know how to use them and will feel comfortable in adopting them. But over time, as happened with printed books, the new media will change the expectations of people.

---

1 See www.rocketbook.com/index.html

2 See www.everybook.net/

3 One company that already has a form of Epaper on the market is E-Ink of Cambridge, MA. Their current electronic paper isn't ready for books, but they have placed some gossamer large advertizing signs in a Mass. JC Penneys, where the text on these featherweight "kites" changes on demand. See www.eink.com/pr_index3.html

4 See www.storage.ibm.com/press/hdd/micro/990617.htm

Let's look at another example of usability—the telephone. What could have been simpler than picking up the handset and telling the voice on the other end to connect you with your mother? The operator knew who you were and who your mother was and completed the transaction. Great usability. Of course there were also good reasons to go to direct-dialed phones—we'd all have to be working as operators today to handle the call volume. So with the advent of the dial, and each of us taking on a second job as our own operator, something had to be done to make it easier for us to remember all these new numbers. In many countries, this was done by associating numbers with each of the ten digits on a dial, and then using those numbers to provide mnemonic memory aids. "800-FLOWERS" is a great example of how businesses have capitalized on this.[5]

It Is Tricky to Push Digits on a Rotary Phone

Of course as our telecommunications network has become more complex, in some ways it has become more difficult to use. Have you ever been in a strange office and had to ask how to get an outside line, or what combination of access codes was necessary to place a long distance call? Come to think of it, how many numbers do you have to tap to get to your long distance carrier from a pay phone, then tap in your credit card number, and finally your destination phone number? We've gone from asking "Ernestine" to place your call, to sometimes having to tap out 36 digits for the same result! If you're calling internationally, it can get even more interesting. Quick—what's the international access code, then the country code. Now, if the published number contains a city code with a leading zero, do you dial that? Sometimes yes, sometimes no, depending on whether you're in-country. Calls placed from different countries are different, yet again. How about the codes for special services, like repeat dialing? Do you remember the *69 code that you can dial to automatically redial a caller when you didn't answer the phone in time?[6] This is "ease of use?"

And then there are our feature-packed pocket cell phones. One of your authors was traveling as he wrote this and thought about storing the number for the airline he was taking into his cell phone's alpha-numeric memory. He thought about it, but never did so, because since he rarely stores numbers, he didn't quite remember the sequence of steps, and it wasn't worth the effort. Instead, he wrote it on a slip of paper where it sat in his pocket, next to the cell phone, until it was needed. Your authors are hardly technophobes, but their cell phones could still use a good dose of "ease of use."

---

5 It's interesting to note that many people believe that it's easier to remember words, such as telephone number mnemonics, than to remember numbers, but in general that's not what's going on. When you translate a bunch of numbers into a word using the letters on a phone dial, you then only have to remember one "memory item," the word, as opposed to 7 or 10 separate numbers. On the other hand, if a group of numbers also represents a single item, such as your birthday, those numbers also condense into a single "memory unit" and are just as easy to remember.

6 We don't understand why the phone companies didn't market mnemonics for these hard to remember codes; they'd probably get a lot more use, and the phone companies would get a lot more revenue.

"More features" in new products can often be a great idea; they can differentiate a product and sometimes, truly, make it easier or more fun to use. But prudent product developers will remember to pay more attention to their intended audience than to their own willingness to memorize "neat" but obscure features. If the average customer can't or won't remember them or how to use them, the product is missing the mark. But if all these neat features are there and easy to use, they may help propel the product toward becoming a winner. Always keep the user firmly in mind. Smart successful designers start this process at the beginning of a product's development not at the end when there's tremendous inertia keeping changes to a minimum.[7]

LET'S TALK! Of course as new technologies develop, they will open up completely new ways for people and machines to interact, and we'll be forced to develop new ideas for ease of use, such as speech recognition. Once our machines become better at understanding random human speech (and they will), how would you want to communicate with a room full of such devices? Eventually, will the machines get smart enough to follow conversation, watch what you're doing, and figure out what you want them to do like a personal assistant, without you having to give them explicit instructions? (If they begin doing this, what if they make the wrong decision and thaw your frozen filet mignon at full power, or sell a stock you really wanted to keep? Will the brokerage firm accept ". . . but it was my APPLIANCE that did it" excuses?) Will you finally be able to record a movie simply by mentioning to your VCR that you want to watch the latest episode of your favorite show this weekend, putting the burden of finding out when it will be broadcast (or downloading it from the Internet) fully on the VCR's shoulders?

Today, in general, we have to give our electronic servants explicit directions for everything we want them to do. We believe that in the not too distant future a growing number of the things around us will pick up some of this burden, reducing the complexity of how we interact with them, and expanding the sphere of what they accomplish for us.

"Usability" is fundamentally about learning, and in the Knowledge Age, we're all lifelong learners. But just as logic designers now build with blocks that contain thousands or millions of transistors rather than worrying about each transistor by itself, our interaction with the things around us will similarly be climbing to a higher level, where we let our silicon servants "sweat the small stuff." We can't wait!

---

7 A good example of a product that makes it easier for a human to use it, rather than trying to train a human to its requirements, is HP's CapShare handheld scanner. Instead of forcing the poor human into scanning ruler-straight swaths of a page, one right next to the other, CapShare lets a person wave the scanner around over the page at random (as with a dust rag) and automatically assembles the entire page correctly. See **www.capshare.hp.com/**

10   What a Beautiful Voice

THAT'S AN UGLY VOICE, IT MUST BE A COMPUTER TALKING   It has been about 20 years since the first computers began to "talk," but nobody would mistake a computer's synthesized voice for that of a person; computer voices have been distinguishably metallic and inhuman, rather ugly. But ever since those first robotic voices appeared, people have been expecting them to get better, to take on human characteristics. Instead, 20 years of immense progress in what computers can do have simply demonstrated how complex it is to create a computerized voice that, in the general case, is indistinguishable from a human. Perhaps that's good thing, or we might begin to wonder if those dinnertime telemarketers are carbon- or silicon-based life forms.

Today from Text to Voice, Tomorrow from Concepts to Expression

One of the reasons for the difficulty is that our voices are constantly changing, with subtle inflections driven by what we want to say, who we're speaking to, and even how we're feeling. To get all of this right, the computer may have to understand the context in which we're speaking and even understand what we're talking about.

We also need to remember that humans have honed "listening" to a fine art. We are exceptionally good at pulling out nuances from the slightest inflection, sometimes even recognizing where a person was born based on vocal cues. These capabilities are, to say the least, difficult to teach to a computer.[1]

Despite these difficulties, computer-generated voices have gotten much better—it's just that they still have a long way to go.

A MULTIMEDIA COMMUNICATION   Of course we don't only communicate with our voices—gestures, a look, even how we position ourselves in relation to the person we're speak-

---

[1] We might find it far easier to teach a computer to bark like a dog, since any nuances in a dog's bark are lost on all but its owner (or perhaps on other dogs.)

ing to communicate volumes. So it's no surprise that it's often easier to understand someone if we can see them at the same time.[2] Instinctively, and without even noticing it, we tend to follow the movement of a speaker's lips,[3] eyes, and other parts of their body. Such cues, for example, help us to decide if a comment is factual or facetious, and if you don't really think this is important, consider how a whole language of "emoticons"[4] has developed to add these missing cues to Email and other typed online communications! They're necessary, and somewhat effective, but not as good as the "real thing" of multimedia interaction. Even though we've come to believe that the telephone, and now Email, is a "natural" method of communication, when the missing "media" *are* available, the fidelity of the communication goes up. Now, thanks to the Internet, some of the missing media are coming back!

For example, if you're trying to describe a complex computer problem, it might be vastly more effective if you could let the service technician on the other end actually view your screen and move your mouse—without flying to your office. There are numerous products available[5] that allow just that, and a growing number of service organizations have recognized that by using such "remote control" software, they can significantly reduce the time required to resolve many user problems. Or consider trying to describe a song to someone by typing about it, compared with sending them the audio file and letting them actually hear it—the extra media make all the difference.

In other cases, using different media for communications is essential. A visually challenged person may not be able to read the text on a screen, but even today's computerized voices can be very effective in conveying the information—just in a different form. Such techniques also apply to specialized jobs where it may not be possible, or safe, for a worker to take their eyes off the task at hand. Similarly, when hearing is not an option, a computer that translates speech into text can be very helpful.

Recognition: Today from Voice to Words and Phrases, Tomorrow from Expressions to Comprehension

In many applications, of course, not one but a combination of media make sense. For example, in searching for a vacation spot online, we might conduct an initial search by typing, then conduct a voice dialog with a customer service person to narrow our choices, and then bring up a video of the hotel or region we're planning to visit. The key is "choice," because

---

2 After childhood, most people find it difficult to pick up another language. Even if an adult reaches a certain level of proficiency with a new language she will still immediately be recognized as a "foreign" speaker by most native speakers. It's not just the vocabulary and inflexion, but all the nuances that are hardwired into a child's brain—even to the point of recognizing and being able to (easily) pronounce certain phonemes (the basic unit of spoken sound). For example, in Italian there is practically no "h" sound, whereas in English this sound is used a great deal; in Arabic there are actually five different "h" sounds, which Americans might find hard to differentiate. The Japanese don't distinguish between the "r" and the "l" sounds, which often makes English speakers smile. Yet the same is true when Italians speak English—an American listener will find it difficult to decide if they're saying "three" or "tree."

3 A research project at the University of Genoa in collaboration with various European research centers is teaching a computer to better recognize phrases and spoken words by learning to read lips. See hera.itc.it:4004/~coianiz/ SpeechReading.html

4 See www.geocities.com/ MotorCity/Pit/4824/smileys.html for an example of common "emoticons."

5 For example, Carbon Copy, see www.compaq.com/products/ networking/software/carboncopy/ index.html or ReachOut at www.stac.com/reachout

each person will value different ways of communicating at different times. Indeed, combinations of media and technologies may come into play with an automated service recognizing that we've exhausted its set of answers and automatically bringing a human travel agent into the discussion, presenting her with a capsule summary of everything we've explored to date. Of course such interactions are very complex.

A simple "what would it cost" question might demand knowledge of the realities of travel that a machine might not (yet) have—the context. So humans, who have an understanding of a broad context, are likely to remain a part of the relationship for quite some time.

HEY, WHAT ARE YOU TALKING ABOUT?

"Context" is so important, and so ingrained in our lives, that it often goes unnoticed, until it's not there. But do you recall the last time you were with an unfamiliar group of people, or in another country where you weren't familiar with their language or customs? It's easy to feel like a fish out of water. The same thing applies to electronic communications, even in familiar environments—the more context that can be provided, the more natural the interaction will feel.

On the other hand, there are good and bad ways of establishing context; take (please!) those infernal "touch-tone trees" we all have to endure with many businesses (and now even with some homes). It sometimes requires four or five levels of choices to establish the context of the interaction you want to have, assuming you haven't hung up in disgust and gone to a different vendor if you have that choice.

One interesting context is that of "language." Although the Web began primarily as an English language medium, by the turn of the millennium the enormous growth of Web pages from other countries will push the percentage of English pages below 50%. Services for translating Web pages to and from your native language will become increasingly important. Already, several companies[6] offer free language translation services online. Although these can be invaluable for getting the gist of the content that's in an unfamiliar language, a quick read of a current machine translation makes it all too apparent that this is one area of computing that still has a long way to go. Today, such translations are broken and often confusing, even though they do (usually) get the point across. But tomorrow, it seems very likely that the day will come when human language transla-

---

6 For example, Alta Vista's free translation service at babelfish.altavista.com/cgi-bin/translate?

tion, text to text or even speech to speech (just like the famous Star Trek translator), will be science fiction no more.[7]

One other bit of context that has generally been missing from our electronic experience is that of the third dimension. With the exception of 3D movies using colored (or the newer, active LCD-shutter) glasses, most of us have to put up with flat 2D simulations of the 3D world—but that too is about to change. Various 3D headsets, some quite good, are coming to market, and one of your authors saw a prototype of a 3D display that didn't require any glasses at all (and it was surprisingly good).[8]

Of course all of these technological marvels do have costs associated with them, and it's reasonable to ask, "Who pays for it?"

---

7 For example, see
www.pcworld.com/cgi-bin/
pcwtoday?ID=10224 and www.
c-star.org/main/english/
noplugin/browser.html

8 See www.compaq.com/rcfoc/
971124.html#Immersive_3-D

Voice Synthesis and Recognition

(COMPUTERIZED) TALK
IS NOT CHEAP

Although talking and listening is easy for us humans, our computers find producing natural human speech, and even more so understanding it, a significant challenge. The Holy Grail of a computer that can do both of these, such as HAL in Kubrick's film *2001, A Space Odyssey,* always *seems* to be just around the corner, but Moore's Law's tremendous gains in processing power have yet to produce the answers to these *HARD* problems. We are still learning about these issues, and the goal remains just over the rainbow.

Homer Dudley, from Bell Labs, is said to have invented the first artificial talking machine in 1936. His "voice coder" or "VODER," as it became known, was primitive by today's standards, but it certainly was a hit at the New York and San Francisco World's Fairs in 1939.

Much of the early speech research was focused on creating, or synthesizing speech, and although this has led to very inexpensive computer programs that can produce understandable speech from text, no one would mistake one of these robotic-sounding voices for a person. Indeed, the difficulty of producing natural speech has surprised many people, along with the fact that success in generating natural speech may require more work on the *other* side of the equation—getting our computers to better *understand* spoken language.

Along the way, universities and industry have produced many experimental, and some commercial speech systems, such as Digital Equipment's (now Compaq's) *DECTalk* hardware,[1] which has been used as a speech aid by the physically challenged noted physicist Stephen Hawking. These days, you don't even need to buy a computer program to translate your text to speech—anyone can type some text into a Web page and be surprised by the quality of the speech that results.[2] When you combine this with far-reaching progress in visual 2D or 3D "talking heads,"[3] even commodity PCs can generate "live" synthesized simulacra that deliver real-time near-human quality speech. These are the basic ingredients of future virtual meeting organizers, virtual sales agents, and more.

So, how does all this happen?

---

1 See www.digital.com/oem/products/dectalk/tdtalk.htm

2 See AT&T's TTS (Text-To-Speech) Web page at: www.research.att.com/projects/tts/ for demos.

3 For example, see Compaq's Cambridge Research Lab's work on facial animation at www.crl.research.digital.com/projects/facial/facial.html Also, an animated head-and-shoulder extension to AT&T's TTS can be seen at www.research.att.com/~osterman/AnimatedHead/

SPEECH SYNTHESIS The first step in creating speech from text is to analyze the written text to be turned into speech just like we did in high school when we had to diagram sentences—each sentence has to be broken down into its component parts, or "linguistic units." Our high school English teachers would be proud!

Only then, when we understand what we have, can we begin to convert the text into speech, and this is a difficult problem. Consider how many ways a sentence can be spoken, each way giving the listener a very different, sometimes even contradictory meaning.[4] For example, "Where are we going?" could be a simple question, a sign of frustration, or just a means to get someone's attention. The culture of both the speaker and the listener also has a lot to do with how a sentence is interpreted.[5]

But if our program has now determined which meaning and inflection is necessary to give to each element of a sentence to be turned into speech, it's not yet finished. It also has to add the gender of the computerized speaker, perhaps give her an accent, and maybe, depending on the circumstances, even alter her voice so, perhaps, she sounds like she has a cold!

Technologies involved

Natural language comprehension, dialogue management, and speech generation in natural language are combined with traditional recognition and synthesis technologies. Starting with spoken utterances, the continuous speech recognition module has the task of recognizing the spoken words; starting with these words then, the comprehension module furnishes a representation of the significance of the sentence. At this point the result can allow for accessing data, interacting with remote systems, and generating sentences in natural language. Finally, the sentences generated in natural language are sent to a vocal text synthesis module, which provides voice responses to the user.

↓ voice

RECOGNITION — Pinpoints the spoken words

↓ words

COMPREHENSION — Understands the meaning of the phrase

↓ meaning

DATA BASE — **DIALOGUE AND TEXT GENERATION** — Meets the user's objective

↓ written phrase

SYNTHESIS — Speaks to the user

↓ voice

---

4 And this gets even more complex in face-to-face conversations. Body language, gestures, and the conversation leading up to each new sentence can alter its meaning. For example, a wink might make it clear we should assume the opposite of what's being said!

5 For example, if you were going to wake up a British girl by knocking on her door the next morning, the way to say this would be "I'll knock you up tomorrow morning." An American girl might interpret this sentence somewhat differently. . .

Nuts and Bolts

In scientific terms, there are three basic methods for generating speech[6]: (1) articulatory synthesizers, (2) formant synthesizers, and (3) concatenative synthesizers. Simply explained, in the first two types the entire conversion process is controlled by a set of "rules." In the third type, very large dictionaries of pre-set speech sounds are used, which, when properly combined, make it possible to generate all the sounds of a given language.

So, at this point, we are constructing sentences out of phonemes (the smallest speech-sound within a language), or out of syllables or sentence fragments, trying to meet the still elusive goal of complete sentences that don't sound robotic or artificial. We're not yet near the end game, but we have made dramatic progress over the past few years.

SPEECH RECOGNITION   Remembering that the intent of this section is to explore how we might relate to the huge number of "smart things" we're likely to have in the Knowledge Age, it's clear that if we're going to interact with them through speech, they have to be able to understand what we say to them, along with being able to speak to us.

Speech recognition[7] in a human involves the ears and the auditory and linguistic centers of the brain. Phonetic, lexical, syntactic, and semantic structures of language are also important in recognizing speech. All of these, plus context and the manner in which the words are pronounced, contribute to our ability to recognize speech.

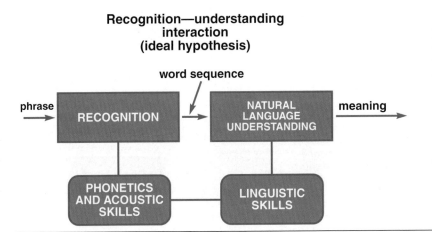

Recognition—understanding interaction (ideal hypothesis)

Phonetic–acoustic knowledge is used during the recognition phase, whereas linguistic knowledge is used in comprehension (e.g., lexicon, semantics, and syntax). However, both are linked to a common dictionary; in the first, the words are characterized by the basic phonetic units that comprise them, whereas in the second, they are characterized by their syntactic category and meaning.

6 See: Survey of the State of the Art in Human Language Technology, sponsored by the National Science Foundation, European Commission: cslu.cse.ogi.edu/HLTsurvey

7 Speech recognition is an extremely complex and multifaceted problem. As an example of how straightforward the understanding of speech is *NOT*, some people, after suffering a stroke, are able to recognize a voice, but not understand what the voice is saying; others understand what is said, but are not able to recognize the voice that said it; still others are able to recognize speech, but are unable to repeat its meaning, either verbally or in writing. These different speech problems throw light on several aspects of human speech recognition, aspects that, in some manner, must also be addressed by our artificial speech recognition systems.

For example, to recognize a single word chosen at random from a dictionary, "acoustic–phonetic" elements are the most important issue; there's little need for "syntactic–semantic" knowledge, which is necessary for the recognition of whole sentences during continuous speech.

As we struggle toward voice recognition in commercial applications such as "directory assistance" systems and voice recognition-based "automated attendant" systems in our workplaces (no more touchtone trees—yes!), we also have to deal with the issue of "speaker dependence"—each of us pronounces the same word differently, sometimes very differently, even within the same language! Differences include the manner of speech (in isolated words, continuous speech, spontaneous speech, etc.), national and regional accents, the extent of each person's vocabulary (more words increase the potential for ambiguity), and the quality of the sound to be recognized (for example, background noise can dramatically reduce understanding for both human and machine listeners, although recent work at the University of Southern California has developed a neural network speech recognition system that is actually far better than a human at picking words out of high background noise[8]).

WHAT DID YOU SAY? No question about it, speech recognition and understanding is your classic "nontrivial task." Researchers are approaching this through both analytical methods and comparative methods that use both prototypes and statistical models.

The meaning of the sentence produced by the comprehension module is interpreted in the context of the preceding sentences by producing a contextual representation of that meaning. This representation is used to access data and to retrieve the required information. The dialogue module must also be capable of generating sentences in natural language to continue the dialogue while providing the information, required parameters and confirmations. A model of the interaction used by the dialogue administrator to coordinate the activities describes the strategy of dialogue management that one wants to pursue. For every application this model must be tested by its users.

How a dialogue system works

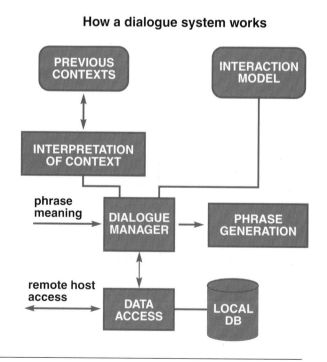

---

8 See **www.usc.edu/ext-relations/news_service/real/real_video.html**

Analytical methods are based on knowledge of the phonetic and linguistic domains, formalizing what is learned into a set of rules that can then be used to interpret spoken utterances. The power of this technology is that, once the acoustic, phonetic, and linguistic knowledge is formalized, it is possible to interpret sentences from any speaker, regardless of their vocabulary. In a certain sense, these methods are similar to languages based on an alphabet. Once the understanding and pronunciation rules are known, it is possible to understand any word, even if we have never heard it before.

Alternately, approaches that compare the *forms* of sounds, rather than codifying the knowledge in advance into a set of rules, tend to start with a collection of examples and then proceed to capture additional acoustic–phonetic knowledge in static form.

Recognition of various speech sounds can occur through comparison with phoneme (basic speech sound) prototypes; these are syllables or words that act as reference models against which an unknown sound or word is compared.

These comparisons cannot be made just for single elements (for example, for one phoneme or syllable), but the program must keep track of, and factor in, many elements, including words and sentences that have already been recognized. This significantly increases the chances for recognition. In effect, recognition becomes easier if a chain is created in which previous recognition helps the recognition of those words that follow. At the same time, those subsequent recognitions reinforce the comprehension of those preceding them! It is with just such two-way chains (called Markovian chains) that spoken language recognition is performed today.

Another methodology, that of comparison (also called "pattern matching"), which was heavily used in the seventies, compares each element of speech with stored prototypes, selecting the most similar prototype as "correct." But this technique has proved more effective for speech recognition systems that have to be trained by each user and so are less effective in programs designed to recognize speech from a general audience.

NEURAL AND FUZZY APPROACHES   Let's look at some of the technologies used to implement speech recognition.

Neural Networks

Neural network architectures, which attempt to emulate the structure of the brain, can be used to perform the comparison operations we described earlier.

Neural networks are made up of many simple parallel-processing elements (analogous to a brain's neurons), interconnected in a way that is similar to the synapses between nerves in a brain. Inputs are processed on the basis of the intensity of signals exchanged between the various "neurons" And as in the brain, the neural network must be trained by supplying it with many examples of phonemes, and by then providing positive reinforcement when its outputs are correct. In this manner the neural network "learns and remembers."

In many cases, "good enough" understanding is sufficient to convey a general meaning.[9] On the other hand, if a radiologist was making your diagnosis and using speech recognition software to dictate his findings to your surgeon, you might prefer that an "exact" rendering of his meaning be conveyed![10]

Fuzzy Logic

For such meaning-critical applications, different techniques are necessary, such as one called "fuzzy logic," which is associated with the mathematical theory of uncertainty. Fuzzy logic applies a rule of philosophy to the context of statistics: Knowledge cannot always be certain. On the contrary, uncertainty and subjectivity play a significant role. These techniques, used in some stock market investment-forecasting programs and to stabilize images in digital cameras, utilize "soft" logic concepts based on common sense, as opposed to "expert systems," which are based on defined rules and probability.

Essentially, a fuzzy logic system might consider that a sound seems similar to (but not exactly like) one it knows, and it will assign the sound a probability that will be considered in light of the probability assignments of the sounds that came before, and come after, this particular sound.

IT'S ALL A MATTER OF SEMANTICS Of course humans are notoriously imprecise in their use of speech, and in how they expect it to be understood. For example, you might ask a speech recognition-enabled car "What time is it?" and be told "It's four forty-five." But you might ask the same question to your human traveling companion and be told "Don't worry, we'll still get there in time; the traffic report shows no backups along our route."

The car's electronics might indeed know your route (from a GPS navigation system), and it might even have access to traffic information (perhaps by "listening" to the radio itself, or by accessing digital traffic information on the Internet), but it would be unlikely to

---

9 Exact meanings aren't always necessary, or even desirable. In some languages, such as Italian, the word by itself does not have a precise meaning, but instead derives its meaning from how it's used in a sentence. In school, Italian children are taught not to repeat the same word, but to use synonyms. On the other hand, English depends far more on the precise definition of individual words.

10 Note that the possible misunderstanding of speech in critical situations is also an issue with human listeners—that's one reason why, in radio communications between air traffic controllers and pilots, aircraft identity and assigned altitudes, which are very critical to all concerned, are spelled out, such as *"Eastern four-five-one—climb to and maintain Flight Level three-five-zero."* The problem of certainty in recognition is also very noticeable in human communication. For example, in the dialogue between control tower and pilot, numbers are always said one at a time in order to avoid confusion; for example, 15 with 50 (15–50) and numbers are repeated by the recipient back to the sender.

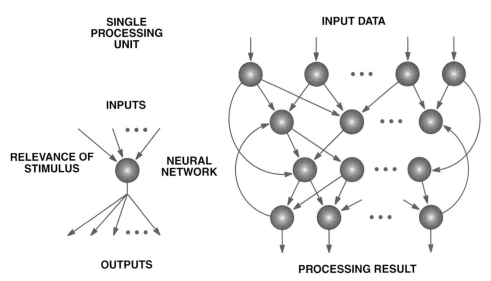

SINGLE PROCESSING UNIT

INPUTS

RELEVANCE OF STIMULUS

NEURAL NETWORK

OUTPUTS

INPUT DATA

PROCESSING RESULT

A neural network is made up of a set of interconnected processing units. Each unit receives signals (inputs) and processes them on the basis of their weight (for example previously established importance, frequency, or input co-presence). On this basis it generates the signals (outputs) to the other units.

"put it all together" to answer what you really wanted to know, as the human did. Semantics and context are major missing elements that prevent most present day speech recognition systems from masquerading as humans (which also defines their future goals.)

Adding "semantic recognition" techniques to basic speech recognition could find applications in many fields, ranging from education to training to entertainment. For example, the development of intelligent agent technologies will be based on intelligent actions and interactions, as well as on their ability to intelligently understand context, which may lead to new uses of "fuzzy" types of recognition. As this work progresses, the day may (eventually) come when our machines not only understand "what we say," but understand the context surrounding our communication to understand "what we mean," which might, eventually, make it difficult for us to determine if we're talking with a machine or a person. That, of course, opens a Pandora's Box of possibilities—and of concerns.

That's a ways off; today our systems struggle to understand just basic speech, much less apply a broader context. For example, a speech recognition program that intended to act on our words (as opposed to "simple" speech-to-text dictation systems) would have to be very clear on the differences in the phrases "leave for Boston" and "leave from Boston." The first implies that Boston is the destination, whereas the second that Boston is the point of departure—simple for you and I, perhaps, but not if we had a silicon heart.

The Markovian Model for vocal recognition. The probability values that connect the points on the graph to the arcs (transitions) (a11 = transition probability b1(k) = emission probability). Every point or status is characterized by the emission of a symbol belonging to the alphabet. The emission is identified by a probability density defined on the possible symbols. During the training phase these probabilities are learned so that the resulting model has the maximum probability of responding in the same manner.

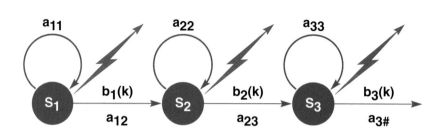

So, a "3D talking head" that not only looks and sounds natural, but also understands and responds to us in a natural manner, is going to require additional work in a wide range of sciences: faster processors with more storage, better techniques in the cognitive and acoustic sciences, and more. But as computers continue to proliferate in all aspects of our business and personal lives, and become ever-smaller and ever-cheaper to the point that they virtually "disappear" into the landscape, we are going to need more natural ways of communicating with them. As we originally suggested, what could be more natural than a kind word.

11   Who Pays?

A GLOBAL VENTURE   The transition from the Industrial Age into the Knowledge Age is the result of many evolutionary paths that are converging in various ways to create a new way of life.[1] "Convergence" isn't being financed by any one industry or consumer group; it is growing in an "organic" way as people figure out how to profitably use the myriad of new products and services that result.

Moreover, initiatives such as the U.S. National Information Infrastructure[2] and the world's G7 Global Information Infrastructure[3] are important accelerating factors and are means to reduce the risks of having a patchwork of "Info-haves" and "Info-have-nots."

"Free" Means Someone Else Is Paying for It

But even as today's and tomorrow's higher speed networks become technically possible, the question that remains is whether people are interested and willing to pay for them. For example, in developed countries, an average family spends about 6% of its income on telecommunications services, and it's unlikely that this percentage would increase in a short time.

Therefore, there's only so much money to support these new networked services, which must be taken into account by those businesses thinking of making the massive investments necessary to bring fast networking to each door. The good news is that broadcast radio and TV models, supported by advertising, are a well-accepted alternative way to pay for such services, and these models have already served the Internet well.

SOME PIECES OF THE PUZZLE   There are basically three components for accessing the Internet. You need a PC (or equivalent "Internet Appliance"), a way to connect to your Internet Service Provider (often a phone line or a special cable TV connection), an account with that Internet Service Provider (ISP). Traditionally, you have to pay for each one. Today, though, we're seeing new

---

1   The convergence of the four "C's" is often at the heart of this: Communications, Computing, Consumer Electronics, and Content. This convergence can be seen in almost every aspect of how we are doing business and living our personal lives. But it isn't the result of a master plan—convergence is resulting from massive innovations, across all of its fields, coming together and reinforcing each other.

2   In the second half of 1997 an initiative began, along with the needed financing, for a high-speed Internet infrastructure to connect universities in North America and beyond. It's commonly known both as "Abilene" and "Internet2." See www. internet2.org/

3   These are research programs financed by the European Community. See www.cordis.lu/info/ frames/if006_en.htm

business models develop to take the sting out of such fees. Some companies will give you a free PC, so long as you either agree to receive advertising on-screen or you agree to a 2- or 3-year contract with an ISP.[4] In the United States, most people can use their existing home telephone to place the modem call to the ISP as a local call, so there is no additional per-minute charge.[5] Although some ISPs do charge a monthly fee for access, a growing number are offering free access, again, in exchange for personal demographic information or for advertising space on your screen. So the question of "who pays for it" gets quite interesting.

By the way, such demographic- and advertiser-supported telecommunications services are not limited to the Internet. Companies such as Callwave and E-Fax[6] offer you free fax service—you sign up, telling them a bit about you, and they give you the use of a regular fax number that you can give out to anyone. Any fax sent to that number gets encapsulated in an Email message and sent to your Internet Email account—for free! This can be a very convenient way to centralize your Email and (previously) separate faxes. Of course these companies may include an ad, or sell mailing lists with your names on them, but that's the trade-off for the free service.

It's clearly getting very difficult for companies to figure out how to charge for their telecommunications services, leading to some seemingly "illogical" pricing.

Today, Distance Doesn't Matter Anymore

For example, as we've all experienced, the distance a call goes no longer relates directly to its cost. Due to pricing structures, we often end up paying more for a long distance call within our state of perhaps 20 miles than we do for a call across the sea to Hawaii. Another example has to do with traditional "leased lines," circuits that businesses would purchase for thousands of dollars per month to carry data a short distance. Today, the advent of DSL (Digital Subscriber Line) services, priced at a fraction of these old circuits' price and often with higher capacity, are playing havoc with established revenue streams. Telecommunications companies can no longer charge a premium based on the type of data being transmitted. Now that everything, from a voice conversation, to a music program, to a TV show, is converted into "numbers" and transmitted over a common network, they cannot differentiate between them and assign a higher "price" to one form of information.

---

4 See www.free-pc.com

5 Note that this is different in most other parts of the world, where people pay a per-minute charge for every local telephone call! But the Internet is forcing this to evolve, most notably in the United Kingdom, which has been leading these changes. First, there were the "free ISPs," which no longer charged a monthly access fee. Users still did have to pay a per-minute local call fee (and the ISPs shared some of that revenue from the phone company). But more recently, some ISPs have been offering the equivalent of "800," or toll-free access numbers (deriving their income from on-screen advertising), so that, for the first time, these users can surf to their heart's content without an eye on the clock. This has resulted in significant growth in online citizens.

6 See www.callwave.com and www.efax.com

Telephone via the Internet

One good example of this is "Internet Telephony." Historically, phone calls were made over the phone company's dedicated voice circuits, which "reserved" a certain path between the caller and receiver for the duration of the call, providing a high-quality "known" connection. Today, though, with the Internet (which will happily carry packets made up of an Email message or a voice), there's no reason that you can't send voice (or video, even) over that same Internet connection! In fact, there are many (some free) programs available that will allow any two people on Internet-connected PCs to spend days on the "phone" without a single long distance charge! Note, though, that because the Internet does not "reserve" a path or bandwidth as does a traditional phone circuit, the quality can be "interesting" and quite variable. This, however, is likely to change as new Internet technologies and protocols, which can "guarantee service," emerge. Phone companies are indeed closely watching these developments, attempting to come up with business models that allow them to continue to add value and make money in this new environment.

Indeed, as basic telecommunications heads toward "the free" (in mid-1999, Sprint just announced 5-cent domestic U.S. phone calls every evening), it is the higher level services,[7] which sit on top of this "basic" telecommunications infrastructure, that may generate the profits. But it isn't necessarily the traditional telecommunications carrier that may be providing them, even though they may add their own value by collecting the best of the services and making them easily available to you.

ACCESS?   Although it escapes many peoples' notice, typically in the United States some of what we pay for local and long distance telephone service has been used to subsidize basic telephone service for rural areas and for some people who could not otherwise afford a phone. With the changing telecommunications markets, this is now becoming more visible as a "Universal Service" surcharge on our bills, but the day may come when this applies to "Universal Internet" service as well. Although access to the Internet is not (yet) classed as a "necessary" utility (as is telephone service or electricity), your authors believe that the Internet is, indeed, the next great "utility," and that such universal access will become every bit as important as access to a telephone.

[7] For example, even though you can easily carry on a videoconference between two PCs over the Internet (if your connection is fast enough), you might be tempted to pay a bit extra for future enhancements. Companies are working on added-value videoconferencing services that would take each individual participant and model them, so they can be virtually placed around a table that every participant sees. Their actual positioning would have nothing to do with their actual location, but the software would make it appear that everyone was sitting at the same table, looking in the "correct" direction to pay attention to another speaker, and even the audio could be modified to seem to be coming from the right place around the table. This enhancement, even though delivered over the "common" Internet, might be a well-appreciated added value.

KEEPING IT WORKING   We've been spoiled. For many years in the United States, if "the phone" stopped working, or if a number didn't go through, we called "The Phone Company" and it was taken care of. Regardless of the (immense) complexity of the network hidden behind our phone, "The Phone Company" was a one-stop shop for resolving the problem. Today, it's not quite so simple.

Is the problem in your home's wiring (which, without a special contract, you have to maintain)? Is it between your home and the local phone company's central office? What if the problem is with your long distance carrier? Troubleshooting problems in such a multivendor environment is often "interesting" and has been known to end up in a case of "finger-pointing" that gets nothing fixed. But if you think that's bad, consider the Internet.

Here, we have a global network made up of hundreds of suppliers with very little formal organization. Add to this the fact that the various packets that make up a single Email message can find their way to their destination across many different vendors' sections of the Internet, and it becomes your classic "nontrivial problem" to point the finger in the right direction. As the Internet indeed becomes an absolute necessity for doing business and living our personal lives, it will have to be able to provide the levels of reliability and availability that we've come to expect from our other "utilities."

Of course there are other issues as well, and, at this time, there are still few answers. But there are lots of opportunities as we continue to enter the Knowledge Age!

"Make It So"

The Knowledge Age is, naturally, focused on and empowered by the ways we create, manipulate, store, and transfer information. It's just the advances we've been discussing in computing power and bandwidth that enable us to do so much with so little (the ethereal "bits" that compose our information assets).

Your authors can easily remember when people asked them what good the new computers were that could only add a few numbers—they couldn't envision how mere numbers, when representing other things and concepts, could take on value far beyond the numbers themselves. And so, we have a thriving economy based on all the things we've learned to do with our favorite bits.

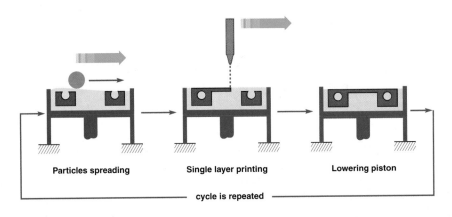

| Particles spreading | Single layer printing | Lowering piston |

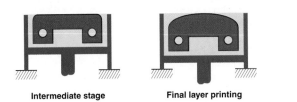

cycle is repeated

| Intermediate stage | Final layer printing | Resulting object |

Three-dimensional printing is a process, currently being developed at the Massachusetts Institute of Technology, for the rapid and flexible construction of objects from a project developed on a computer screen. The working principle is based on the creation of strips of the object. The process begins by distributing first a thin strip of granules on a base. By using a technology similar to that of the ink jet printer, a binding material is then injected that binds the granules to each other. The process is repeated after each strip is laid down, until the object is completed.

For example, see http://me.mit.edu/groups/tdp/

BACK TO ATOMS!   But as much as we can do with bits, there does come a time when we need physical goods and services—when bits alone are not enough. We can browse an online catalog of beautiful china, but it's very hard to eat off our monitors— if we like the china enough, we probably want to have it physically on our table. Similarly, taking a virtual vacation through the Web can help us decide where we might like to go, but we won't get a tan or feel the sand under our feet unless we get on a plane (at least not today).[1] We still need to integrate the world of bits with the world of atoms.

There are some technologies, such as StereoLithography, that can actually take a computer-generated 3D model of something and "make it so," converting the bits of its pattern into a real physical item.[2] Other techniques for "making it so" are being developed at MIT, where a modified ink jet printer prints bits of "filler," one tiny strip on top of another, to actually create the 3D object modeled in the computer.[3]

Although such techniques have great value, there are many more applications where manual intervention is still required. And that manual intervention can be provided by people, or by robots.[4]

ROBOTS CAN DO IT   Of course the science fiction robots of books and movies are still (happily) a long way away, but a growing class of industrial, and perhaps soon domestic robots, may be able to help tie the virtual world of information with the physical world in which we live.

On the domestic scene, Sony recently completely sold out of its first batch of robot dogs, called "Aibo"[5] in a few hours, even though Aibo is still in many ways an expensive "toy." But other domestic robots, such as "Cye" from Personal Robots,[6] may (eventually) take up the task of vacuuming the house!

Of course those robots don't "make" things; yet industrial robots have been helping to make the things we take for granted for years.

Some of these robots extend a person's reach, either over distance or into inhospitable environments (such as inside a nuclear reactor); they can extend our reach through adding

---

1 The transporter, so commonly used on Star Trek, may go beyond not only current technological capacity, but also beyond the laws of physics. For interesting observations on what may be feasible, see *The Physics of Star Trek,* L. M. Krauss, Harper Perennial, 1996. Even if "teleportation" does become feasible, it might bring with it some interesting psychological issues—see *The Minds I,* D. R. Hofstadter, Bantam Books, 1985.

2 See www.advproto.com/sla.htm

3 See web.mit.edu/afs/athena.mit.edu/org/t/tdp/www

4 The word "robot" derives from the Czech word *robota* (to work) and was used for the first time by the writer Karel Capek in 1920 in a science fiction play to identify automatons who worked in place of human workers.

5 See www.world.sony.com/robot/top.html

6 See www.personalrobots.com

strength or by helping us to work on tiny structures, such as on the retina in an eye, where our "huge clumsy" fingers would do a poor job. We might call this class of robots, "telepresence." But telepresence robots have some interesting challenges. For example, if you were sitting in Boston and wanted to use a telepresence robot in Kyoto, Japan to do your rice paper painting, there's far more to it than just having that distant robot mimic your hand movements. You'd have to be able to "feel" how hard the robot was holding the brush, because that will affect its stroke, and as you crumble the dry ink, you'd have to be able to sense the consistency of the powder. Similarly, an eye surgeon finds great value in being able to sense the differing feel of the various layers of cells she is cutting through. (Indeed, a robot could augment a surgeon's senses, assigning a "unique" feel to different tissues that a regular hand might never be able to sense.) Also on the medical scene, work is going on to enable surgeons located far from a patient to perform surgery through such telepresence robots, bringing expertise to areas where it may not otherwise be available. Of course there are a few issues—you wouldn't want the computer to crash or the connection to be dropped at a critical point.

Another type of robot is more autonomous, operating in its (typically limited) environment based on information about the world around it that it gathers itself. These robots, for example, may be welding sections of a car together on an assembly line, or making the chocolates that you buy in the candy store. In fact, in some Japanese factories, there are robots making other robots!

But the idea of robots is not limited to our classic view—as technology continues to grow and as information about the world around us, including our roads and traffic conditions, continues to become more available, we may see robotic trucks[7] that extend the output of these robotic factories right to our doors!

At the other end of the spectrum, we may be seeing robots wandering around in our bodies performing surgery that a human doctor could never imagine. "Nanomachines,"[8] incredibly tiny machines built on the molecular scale, are being designed that will be able to, quite literally, go where no one has gone before, such as into blood vessels. Future tiny "Roto Rooter" nanomachines may clear our blood vessels as easily and painlessly as a plumber might clear our drain pipes and other nanomachines may sense body chemistry and automatically release just the right amount of a drug when it's needed (as in controlling a diabetic's insulin level). Nanomachines are not only future devices—the sensors that deploy our cars' airbags are some of the first mass-produced nanomachines, and tiny computer-controlled mirrors are being used to switch optical communications signals.

---

7 A robot truck with these abilities was presented at the conference "Advancing the Information Society" in Barcelona on February 5, 1998, where the point was made based on results of the Telematics research programs of the European Community (www.echo.lu/telematics).

8 See nano.xerox.com/nano and mems.engr.wisc.edu/what

The idea of turning insubstantial bits into real products made of atoms is really nothing new to any of us with a printer connected to our PCs. When you hit the "print" button, that's just what you're doing—converting bits into atoms. Even though some books are available to download onto your PC, and you could then print them, it would be a slow process, and the result wouldn't be nearly as convenient as a regular book. Borders, the bookseller, and a company called Sprout, are out to change this. They will be putting a kiosk in Borders' bookstores (and later, perhaps in non-bookstores) that will have the ability to rapidly download and print a book that might not otherwise be in stock. You put in your order and one section of the machine prints the black and white pages while another section turns out the color cover for what will become your very own instant paperback.[9] Again, bits into atoms.

So as we can see, robots are an important element in our ability to translate information into reality, and they may, over time, become so widespread that they will change the relationship between people and things. But will they ever develop the personalities, benign or threatening, of their science fiction predecessors? We will have to wait and see.[10]

The photos emphasize the degree of miniaturization attained in the production of electromechanical components, in this case gears used in the construction of an electric motor, placed on the head of an ant, near its legs, and on the leg of a cockroach.

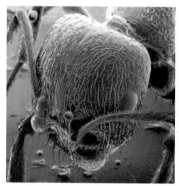
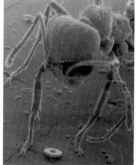

---

9 See www.pioneerplanet.com/search?NS-search-page=document&NS-rel-doc-name=/docs/head11.htm&NS-query=sprout+and+borders&NS-search-type=NS-BOOLEAN-QUERY&NS-collection=PioneerPlanet&NS-docs-matched=14&NS-doc-number=1

10 One group exploring these "things robots" is "The Robot Group," online at www.robotgroup.org

12　It Doesn't End Here . . .

If all the changes we've been discussing seem to be just "too much," "too fast," and you're thinking that a "kinder, gentler pace of change" might be nice—we hate to disappoint you. Every indication, at this point, is that the rate of innovation is not only going to keep moving quickly, but will actually accelerate! Moore's Law[1] has been choreographing the dance of our computers for over two decades, and this has changed almost every facet of our lives.

Yet in some fields, Moore's Law itself is being eclipsed—for example in the storage capacity of disk drives. In a recent speech,[2] SGI Chief Scientist John Mashey pointed out that, historically, disk drives have been increasing their density 1.3 times per year. However, since 1990, that rate of storage improvement reached 1.6 times per year, and since 1998, disk drives are improving their price/performance two times every year—faster than Moore's Law.[3]

If we look at some of the revolutionary changes that may affect the very basics of how our future computer chips work (such as DNA computers, chemically driven self-assembling molecular computers, and quantum computing),[4] everything we've seen to date could be the barest appetizer of this meal!

Bandwidth is burgeoning as well. In 1999, a single state-of-the-art fiber could carry not one, but 160 full 10 gigabits per second data streams (each stream a different wavelength, or "color" within the fiber); that's 1.6 terabits per second per hair-thin fiber. New advances promise to raise the number of "colors" that can be used on a single fiber from 160 to 2,000—and to carry that 20 terabits per second of data as far as 2,000 miles without having to amplify (regenerate) the signals![5] (And don't forget that hundreds of fibers can be combined into each bundle, so the data capacity of even a single bundle of fibers will be beyond comprehension.) Even the "last mile"—getting data to our homes and offices and even to our pockets—is slowly getting faster. Cable TV modems and DSL

1 Gordon Moore, one of the founders of Intel, predicted in the mid-1970s that the number of transistors on a chip would double every 2 years while the chip's cost remained the same. Some years later he revised this to every 18 months. And he has been right ever since.

2 1999 USENIX speech "*Big Data and the Next Wave of InfraStress.*"

3 See www.compaq.com/rcfoc/ 19990802.html#Our_Incredibly_ Shrinking for additional examples.

4 See www.compaq.com/rcfoc/ 19990726.html#Weve_Only_Just for some details.

5 July 1999 Gilder Technology Report (www.gildertech.com/).

services offer tens to hundreds of times faster service than a traditional modem, and the next generation of pocket cell phones may support 2 megabits per second!

In light of these incredible changes, we'd like to summarize some of the products and services that we have explored so far, but we'll look at them from the perspective of what still must be done to make them pervasive and widely available. In fact, it is only when the fruits of the telecommunications, computing, content, and consumer electronics worlds "converge," when they "disappear" and become an integral and natural part of our world, that the Knowledge age will have truly arrived. But remember that this is just a snapshot and just a beginning.

Research taking place today will progress in new "organic" directions, branching off into new paths as a result of each new discovery. We will all have to be agile and quick to respond if we hope to profit from the results.

We expect that the evolution of the products and services we're just beginning to use today will lead to a world that, in many ways, becomes distance insensitive. "Telepresence," the use of rich multimedia, and pervasive and inexpensive access will make it easy to not only share information at a whim, but to engage in distant interpersonal relationships far richer than those supported by the "mere" telephone.

According to the Visionary Group,[6] new products and services that are likely to develop can be grouped into several areas, each with challenges that require significant investment in research:

The Storage and Intelligent Access of Multimedia Information

The artificial distinctions that our computers have made between different forms of information (text, voice, video, etc.) will be erased, as a growing number of devices seamlessly handle the spectrum of media. This will open some interesting services, such as these which you might request from an (automated) online video store:

- I would like to see video Number 8 from the Sports catalog.

- I would like to see the latest James Bond movie.

- I would like to get a short film suitable for my 12-year-old grandson, who likes adventures with a happy ending.

- I would like to see a documentary on Native Americans from their perspective.

6 *Communications for Society,* February 1997, Visionary Research, European Commission, Office for official publications of the European Communities, CD-98-96-994-EN-C.

- I would like to see some movies that support the theories of an international plot to assassinate J.F.K.

- I would like to see that movie where she gives him a long kiss.

Today, of course, we're pretty much limited to the first request, although some others are possible with great difficulty.

Another issue is the vast amount of multimedia information—how do we figure out what's important to us? Today, for text information, we can use filters and search engines, but we don't yet have the technology to broadly analyze audio and video streams to hunt for the "headlines" that may interest us (editors are still very much in demand).

TELEPRESENCE IN THE GLOBAL VILLAGE   We're talking about virtual meetings attended by real people, the ability to visit far-away places without having to physically go there, gaining new experiences through simulations, participating in virtual "parties" in a way that gives us the social experiences of real ones—these are just some of the interesting possibilities that truly rich at-a-distance interaction technologies may bring.

Today, videoconferences are a poor substitute for a face-to-face meeting, but extensions to today's technologies, which will improve definition and provide 3D video and audio positioning, hold the potential to make these experiences more "real."

For example, 3D sound coupled with treating each individual in the conference as an "object," can give us many nonvisual cues as to where someone is located and even which way their head may be turned.

From the visual perspective, artificial 3D environments will continue to improve, as will "immersive" devices that provide a 3D first-person perspective. The combining of simulated and mapped environments, plus possibly input from satellites and even from remotely controlled flying or floating "WebCams," may eventually enable us to visit almost anyplace on (or off) Earth without ever earning a Frequent Flyer Mile. And this technology could easily extend to new forms of "chat rooms," making distant person-to-person interaction far more rich. Who knows, you might dress your virtual representative up "to the nines" in a virtual costume to attend a virtual ball.

VIRTUAL PEOPLE Of course it won't really be "us" inhabiting these virtual environments, but our virtual selves (avatars, simulacra, etc.). The avatar would appear to others to be "us" (or what we wish to project as "us"), and we might directly control it (using "Data Gloves" or other 3D controllers or, hopefully, less constrictive and invasive interfaces, perhaps based on machine-vision systems watching us). Or, we could allow our avatars to interact with the environment in a semi-autonomous manner.

We've already seen the beginnings of the "semi-autonomous" avatar in the guise of computer-generated opponents in computer role-playing games, and within the game's limited environment, they can be effective opponents indeed. As this technology improves, so will the scope in which our avatars might operate, and we could end up using such "representations of people" for some very un-game-like activities, such as gauging shoppers' reactions to store layouts, or watching how a crowd of avatars deals with an emergency stairway during a fire.

We're also seeing the very beginnings of autonomous agents that can live on the Net and do our bidding, such as keeping track of low travel fares to an interesting destination or flagging us when mortgage interest rates fall to the point we should refinance our homes. But this could go much farther. How about an audio/video agent that answers phone or videoconference calls and has some limited ability to interact with callers in your stead? They could give a specific message, depending on who called, and they might also be able to negotiate the purchase price of something we're interested in buying.[7] Or, could a more sophisticated agent detect a telemarketer's call and then engage them in a long (and fruitless) conversation? Now that agent would sell really well!

Another use of virtual personalities might be for health care. Of course this is very much a double-edged sword, with (conceivably) high-quality diagnostic information available to anyone at any time, but with the specter of incorrect or incomplete diagnoses—especially if a person wasn't sure if he or she were interacting with a software construct or a real human health care worker.[8] But with (quite a few) advances, today's first steps in computer diagnosis and even remote surgery (being explored by the military for battlefield injuries) hold the long-term potential for changing medicine.

7 The MPEG4 standard already incorporates the concept of Intelligent Agents for interacting with different suppliers around price.

8 Do you think that's impossible? Then consider Kyoko Date. She's an 18-year-old, 97 pound Japanese teenager, and a rock star with successful music videos selling in Japan, and she is the heartthrob of many a Japanese teenage boy. But she's not real. Although she makes money and has global fan clubs, Kyoko Date is a software construct, and as technology improves, we may see her interacting, live, with talk show hosts and perhaps with people on the Internet! See a Quick Time movie fragment of Kyoko Date at home.inreach.com/macbain/misc/DK96.ZIP and read an interview with her at www.dhw.co.jp/horipro/talent/DK96/int_e.html Additional information on Kyoko is at www.dhw.co.jp/horipro/talent/DK96/index_e.html

PARTNERSHIP BETWEEN PEOPLE AND MACHINES When the automobile first came out, you had to be an expert, or at least a tinkerer, to drive them. Just puttering down to the local store could be a challenge laced with tools and dirt under your fingernails. Sounds a bit like using computers today—and for similar reasons. The computer is really still very immature. To use it for other than "canned" tasks, it requires too much of an investment in time and too much technical expertise to keep it all working and updated. But as the car matured to something that (usually) "just works," so too will our computers improve. One day, instead of having to pedantically tell our computers each and every thing we want them to do (as we would with a tool such as a hammer), they'll transform into "assistants" that will learn from us; they'll remember, and take some initiative.

Say It

Of course the traditional keyboard and mouse might not be the ideal interface with such an assistant. Natural spoken language would seem, on the surface, to be an obvious way to interface with a computerized assistant, much the way we might interact with a human personal assistant. Indeed, in many cases speech might work well. On the other hand, would you want to treat your computerized refrigerator or microwave as a person? If these things did understand your speech, what might they do if you became 'impolite'? Would the refrigerator melt your ice cream and your microwave burn your muffin?

The Face of the Machine

How might you want your computerized assistant to appear? Although some people formed a less than positive impression of early animated on-screen "helpers," more sophisticated entities might be useful as a focal point for our attention. If these "helpers" could make use of computerized vision systems to actually watch us and use voice analysis systems to listen to us, they might be able to follow our interest and emotional state and adjust the information they're providing accordingly!

Indeed, the types of devices that we use to interact with the future information world are likely to be different from what we use today. For example, today's most common electronic interface, the telephone, is woefully inadequate for the demands of the Knowledge Age. (Even with software that let's you listen to your Email over the phone, or surf a Web page, a telephone interface is (if we're kind), difficult. The phone's only advantage is that it's just about everywhere.) But beyond phones-with-screens, we're likely to see some more radical changes. For example, suppose your laser printer could "print" a tiny active tag onto each sheet of paper as it's printed. This radio-accessible tag would uniquely identify the page, and every work surface in your office might be "activated" to notice what sheets were lying on them. Suddenly, a request (perhaps through a touch screen) for information that happens to be on a piece of paper could be satisfied by your search engine! Not to mention that such an office would be able to find everything in it.[9]

Similarly, such an active desk, combined with the 3D video and audio processing techniques we discussed earlier, could be virtually extended to group meetings, with each member "inserted" into whatever meeting venue was appropriate.

It's Everywhere!

Of course most of us don't always stay at our desks, and as the explosion of mobile phones, pagers, notebook computers, and PDAs has shown, people want to be increasingly in touch with their information wherever they may be. This demands that the wireless infrastructure continue to improve, and the devices that interact with it become even less obtrusive. For example, one of your authors purchased a Motorola Star Tac phone shortly after it came on the market. This phone was so small and light compared to others available at that time, that for the first time it became "invisible." It fit in a shirt pocket with nary a bulge, and it was so light that his biggest fear was that he'd leave it in the pocket when the shirt went into the wash! As miniaturization continues, even to the (eventual) point where circuits can be woven into fabric itself, we'll find ourselves in a very connected world. On the other hand, all these devices had BEST come with an "off" switch!

Touch It

But still, there are many applications where we need an even more intimate interaction with the world of data. Take, for example, remote surgery. The surgeon clearly needs to have an excellent 3D view into the patient (literally), but she also needs to make use of additional senses, such as hearing

9 This concept is far from science fiction. Sony announced a form of intelligent label in January 1998 to be put on videocassettes; the VCR writes content information, and the lengths of segments, into the tag, which can then be accessed later without fast-forwarding the tape. Assumedly, the tags could also be interrogated by a computer to store the information into a database of all your video clips.

the sounds that the patient (or her tools) are making, and she needs tactile feedback when her remote fingers and tools touch tissue. Even if you're not a surgeon, if you've tried a flight or driving simulator, both with and without "force feedback," you'll instantly understand its value.[10]

Entertainment

By the way, all of these technologies taken together, and removed from the button-down work environment, also hold the potential to provide for highly engaging interactive entertainment experiences. Indeed, if we look back at many of the multimedia technologies we take for granted in today's workplace, they directly resulted from advances in computer gaming. Given peoples' interest in entertainment, this could be a tremendous market.

Personalized for You

But whatever capabilities and interfaces our future devices sport, we can be pretty sure that we'll be able to customize them in any way that makes sense (and many ways that won't make sense to some of us.) As we acquire more of our information and interaction through communicating computing devices, we're going to want to feel increasingly comfortable interacting with (and through) them. So from today's phones that can play whatever we want as a "ring" sound, to perhaps tomorrow's intelligent agents that look, sound, and feel just as we'd like them, we can expect a very personally customized world.

RIGHTS AND DUTIES   The Knowledge Age doesn't, by itself, change anything relating to what we owe our societies or the rights those societies confer upon us. But the Internet represents a major acceleration of the process, begun with the printing press, of putting information directly into "ordinary" peoples' hands. That will lead to significant changes. For example, the Knowledge Age could, conceivably, lead to direct-democracy forms of government (where every citizen could vote on every issue of interest to them, rather than by having such votes go through elected representatives). But at the same time, such easy and pervasive access to information raises very real privacy issues.

For example, today, a few "famous" people suffer the fishbowl effect of having the paparazzi follow their every move and broadcast their lives to the world. But if some of the technologies of the Knowledge Age, such

10 In August 1999, Logitech announced "FEELit," a $99 mouse that adds the sensation of tactile feedback as you move the cursor across the screen. In effect, this adds the dimension of physically interacting with on-screen objects, such as being able to feel a rough texture, or sense a viscous area that it's hard to move across. See www.wired.com/news/news/email/explode-infobeat/technology/story/21207.html

as WebCams and closely tracked cell phones, were to be used in inappropriate ways, any of us could experience the downside of having the world as our constant companion. Even without such devices, the information trail we can leave in cyberspace can give an interested information collector a picture of our movements and interests that we might not wish to expose!

For example, with PCs connected to the Internet, it's not hard (nor unknown) for manufacturers to keep notes on who has what versions of their software installed. Some Web sites leave "cookie crumb" trails of our activity hidden away in "cookie" files on our hard disks, which remind them what we've been up to on our next visit.

Not all of this is bad. The information retained in cookie files can make our Web surfing experience smoother and more productive. But it does highlight the increasingly subtle differences between intrusive invasion and sharing, which we'll have to address as new techniques are developed and as the boundaries between our private systems and the extended Internet continue to become less clear.

This global interactive environment was never planned. Its evolution, from the halls of research and education into a dramatic economic and social force, was not designed with an end goal in mind, so strong security capabilities, such as authentication, encryption, guarantees of anonymity (where appropriate), or even the ability to recall a nasty word were never designed in, but they are becoming increasingly important. The good news is that, technically, these issues can all be addressed. We only have to recognize their importance and ensure that we protect our rights as we emigrate to this new land of cyberspace. The choice is very much ours.

ELECTRONIC MARKET It seems undeniable that the global electronic marketplace is destined to make a dramatic economic impact on our world, changing the ways and the paths through which people do businesses. (For example, the rise in online auctions between individuals reduces some commerce between individuals and traditional businesses.) As Ecommerce continues to unfold, everything will have to adapt: the payment methods we use; guarantees that are made; cross-border regulations; even tax issues, must evolve to keep up with these new ways of doing business.

VIRTUAL AND DISTRIBUTED ENTERPRISES Business will never be the same. Some people no longer have to commute and crowd into traditional offices, potentially letting them work closer to their customers and offering enhanced quality of life. Businesses both compete and cooperate at the same time (coopetition), and may tie their information systems together over the Internet to enable virtual teams to spring up, do a project, and dissolve. Borders, shores, and time zones not only cease to matter but can be leveraged through "follow the sun" techniques to use a global team to its best advantage. A large number of new services supporting such ventures will open opportunities for many businesses.

Not the End, Only the Beginning

This journey has not been about "answers." Instead, we've tried to describe some of the things that are changing right now, and to explore some changes that may yet occur. We hope you'll reflect on them as a vehicle to think about how different things already are, and how much more different they are going to become—and so to consider the vast opportunities that exist for those of us willing to embrace the changes. Conversely, it's also a warning for anyone struggling to keep the lid of Pandora's Box closed with all their might—that activity will probably exhaust them, and they'll become road kill by the side of the Information Highway.

But we also need to realize that the way the Knowledge Age will continue to unfold is hardly cast in stone. Individual and global societies may embrace (or not) aspects of what's happening today in ways we can't yet foresee. Of course revolutionary technology changes could make everything we've experienced to date moot. It's only by keeping up on the constant changes in technology, and in business and society, that we can even hope to chart the best course for our businesses and for our personal lives.

Will the egalitarian aspect of the Internet—the empowering of the "little guy" that we've recently seen—prevail? Or will the monolithic telecommunications companies again extend monopolistic control over how we communicate and share information? Although your authors suspect that we will never go back to the old model, there are no guarantees.

But one thing does seem clear to us. As computers get more powerful, as bandwidth becomes more available, and as access to the resulting information utility becomes more pervasive, a vast number of communicating computing appliances, which will bear little resemblance to those we use today, will spread across our landscape, and they'll become so ubiquitous, so common, that the day will come when we fail to notice they're there at all; they'll "disappear." At which point they will have truly changed how we work, live, and play.

It's up to each of us to ensure that this happens in a way we can, quite literally, live with.

We've enjoyed taking this journey with you.

Index

Page numbers in *italics* indicate figures.

About the Authors

Roberto Saracco has led research activities in Telecommunications Management at CSELT—the Telecom Italia Group Research Center (www.cselt.it) since the mid-1980s. His position at CSELT, which was at the forefront of research in telecommunications and information technology, stimulated his interest in the evolution of the Information Society.

Since 1994 Mr. Saracco has been head of the Marketing & Communications Division at CSELT, where he has dedicated considerable effort to spreading an understanding of technology's role in the evolution of society. His work in the marketing area takes him around the world, providing a wonderful opportunity to share ideas and absorb different perspectives.

Mr. Saracco has participated in international standardization organizations that have included CCITT, OSI, ETSI, and T1M1. A member of the advisory board of the *Journal on Network and Systems Management* and an active member of the IEEE, Mr. Saracco chaired the Committee on Network Operations & Management from 1994 to 1996 and the Committee on Enterprise Networking from 1996 to 1998. Currently, he is the Secretary of the Communications Society's (COMSOC) Technical Affairs Council. From 1996 to 1997, he chaired the Visionary Group on Super Intelligent Networks with the objective of steering cooperative research at EU level beyond the year 2000. Mr. Saracco has published many technical and popular papers and a number of books.

Jeffrey R. Harrow graduated from the University of South Florida, but his passion for technology began long before when he and a group of other teenage ham radio operators implemented one of the first touch-tone mobile telephone systems. He has been chief engineer of a commercial broadcast FM station, an engineer for a two-way mobile telephone and paging company, a commercial pilot, an instrument flight instructor, and a radio announcer.

Mr. Harrow worked for Digital Equipment Corporation (DEC) to implement a network design and installation business in the southeast U.S., and subsequently moved into a corporate services position working on network products and services.

Currently, Mr. Harrow works as a senior consulting engineer for Compaq's Technology & Corporate Development Organization where, through consulting and seminars, he helps people worldwide understand and profit from the innovations and trends of contemporary computing. In addition, he writes the weekly technology journal, the *Rapidly Changing Face of Computing,* which is available on the Web at http://www.compaq.com/rcfoc.

Robert Weihmayer joined Bell-Northern Research early in his career to develop network planning tools. He worked on a series of systems based on engineering costing models and combinatorial optimization algorithms for optimal transition planning to digital switching and transmission systems. This work was continued in GTE Government Systems and GTE Laboratories with optimization tools for logistics and deployment planning, as well as data network evolution planning systems.

After joining an Artificial Intelligence (AI) Research Department at GTE Laboratories, Mr. Weihmayer became principal investigator of the Distributed AI Project. This led to many experiments and papers dealing with distributed cooperative agents jointly solving network operations problems. He then became head of the Intelligent Systems Department at GTE Laboratories. He is currently Director of Information Technology, Enterprise Systems, with GTE Internetworking.

Mr. Weihmayer has published over 20 papers in all the technical areas previously mentioned. As an IEEE member, he has worked in organizational capacities in many IEEE workshops and conferences. He is currently technical cochairman of NOMS 2000—the IEEE Network Operations and Management Symposium.